2017 SQA Specimen and Past Papers with Answers

National 5
GEOGRAPHY

X SQA

2015 & 2017 Exams
and 2017 Specimen Question Paper

HODDER
GIBSON
AN HACHETTE UK COMPANY

This book contains the official SQA 2015 and 2017 Exams, and the 2017 Specimen Question Paper for National 5 Geography, with associated SQA-approved answers modified from the official marking instructions that accompany the paper.

In addition the book contains study skills advice. This advice has been specially commissioned by Hodder Gibson, and has been written by experienced senior teachers and examiners in line with the new National 5 syllabus and assessment outlines. This is not SQA material but has been devised to provide further guidance for National 5 examinations.

Hodder Gibson is grateful to the copyright holders for permission to use their material. Every effort has been made to trace the copyright holders and to obtain their permission for the use of copyright material. Hodder Gibson will be happy to receive information allowing us to rectify any error or omission in future editions.

Permission has been sought from all relevant copyright holders and Hodder Gibson is grateful for the use of the following:

Image © Gary Whitton/Shutterstock.com (2015 page 10);
Image © Kanokratnok/Shutterstock.com (2015 page 10);
Image © markobe/Fotolia (2015 page 11);
Image © Ivan Cholakov/Shutterstock.com (2015 page 12);
Image © Lledo/Shutterstock.com (2015 page 12);
Image © De Visu/Shutterstock.com (2017 page 7);
Image © Fabio Lamanna/Shutterstock.com (2017 page 11);
Image © Richard Roscoe, Photovolcanica (2017 SQP page 15);
Image is reproduced by kind permission of the Fairtrade Foundation © David Macharia (2017 SQP page 16);
Image © jan kranendonk/Shutterstock.com (2017 SQP page 17);
Ordnance Survey maps © Crown Copyright 2017. Ordnance Survey 100047450.

Hachette UK's policy is to use papers that are natural, renewable and recyclable products and made from wood grown in sustainable forests. The logging and manufacturing processes are expected to conform to the environmental regulations of the country of origin.

Orders: please contact Bookpoint Ltd, 130 Park Drive, Milton Park, Abingdon, Oxon OX14 4SE. Telephone: (44) 01235 827720. Fax: (44) 01235 400454. Lines are open 9.00–5.00, Monday to Saturday, with a 24-hour message answering service. Visit our website at www.hoddereducation.co.uk. Hodder Gibson can be contacted direct on: Tel: 0141 333 4650; Fax: 0141 404 8188; email: hoddergibson@hodder.co.uk

This collection first published in 2017 by
Hodder Gibson, an imprint of Hodder Education,
An Hachette UK Company
211 St Vincent Street
Glasgow G2 5QY

Typeset by Aptara, Inc.

Printed in the UK

A catalogue record for this title is available from the British Library

ISBN: 978-1-5104-2186-8

2 1

2018 2017

Introduction

National 5 Geography

This book of SQA past papers contains the question papers used in the 2015* and 2017 exams (with the answers at the back of the book). The National 5 Geography exam is being extended by 20 marks from 2018 onwards, following the removal of unit assessments from the course. A new specimen question paper, which reflects the requirements of the extended exam, is also included. The specimen question paper reflects the content and duration of the exam in 2018. All of the question papers included in the book (2015, 2017 and the new specimen question paper) provide excellent practice for the final exams.

Using the 2015 and 2017 past papers as part of your revision will help you to develop the vital skills and techniques needed for the exam, and will help you to identify any knowledge gaps you may have.

* Questions from the 2016 past paper have been used to create the new specimen question paper. To avoid duplication and provide you with optimum variety of questions, we have intentionally included the 2015 past paper instead.

The exam

The course assessment will consist of two parts: a question paper (80 marks) and an assignment (20 marks). The question paper is therefore worth four-fifths of the overall marks of the course assessment, and the assignment one-fifth. The assignment is completed throughout the year and submitted to SQA to be marked around April. This is worth 20 marks. These marks are then added to the marks achieved in the exam paper to give you a final award.

The question paper

The purpose of the question paper is to allow you to demonstrate the skills you have acquired and to reveal the knowledge and understanding you have gained from the topics studied throughout the course. The question paper will give you the chance to show your ability in describing, explaining, matching and evaluating a broad range of geographical information as well as using a variety of maps and demonstrating proficiency in Ordnance Survey (OS) skills. Candidates will complete this question paper in 2 hours and 20 minutes. Questions will be asked on a local, regional and global scale. The question paper has three sections.

Section 1: Physical Environments

This section is worth 30 marks. Candidates will answer a mixture of limited/extended-response questions by using the knowledge, understanding and skills learned throughout the course. In this section there is a choice. Candidates should answer **either** Question 1: glaciation/coasts or Question 2: rivers/limestone. This will be dependent on the subjects taught at your school. Some topics you could be asked to answer questions on include **Weather, Landscape formations** within Scotland and/or the UK, and **Land use management** – conflicts and solutions. In this section you may also be examined on Ordnance Survey skills using a map.

Section 2: Human Environments

This section is worth 30 marks. Candidates will answer a mixture of limited/extended-response questions by using the knowledge, understanding and skills learned throughout the course. Candidates should answer all questions in this section. Questions in this section are drawn from both the developed and developing world. Some topics you could be asked questions on include **Population** (development indicators, population distribution, factors affecting birth rates and death rates), **Urban** (land use characteristics in cities in the developed world, recent developments in developed world cities, strategies to improve shanty towns) and **Rural** (changes in rural landscapes in both the developed and developing world). In this section you may also be examined on Ordnance Survey skills using a map.

Section 3: Global Issues

This section is worth 20 marks, made up of two 10 mark questions. Candidates will answer a mixture of limited/extended-response questions by using the knowledge, understanding and skills learned throughout the course. In this section there is a choice of questions. Candidates should answer **two** questions from a choice of six. Your choice will be dependent on the topics taught at your school. The choice of topics is: **Climate change**, **Natural regions**, **Environmental hazards**, **Trade and globalisation**, **Tourism** and **Health**.

Types of questions

The main types of questions used in the paper are: Describe, Explain, Give reasons, Match, Give advantages and/or disadvantages, and Give map evidence.

Describe questions

You must make a number of relevant, factual points. These should be key points taken from a given source, for example a map, diagram or table.

Explain or Give reasons questions

You should make a number of points giving clear reasons for a given situation. The command word "explain" will be used when you are asked to demonstrate knowledge and understanding. Sometimes the command words "give reasons" may be used as an alternative to "explain".

Match questions

You are asked to match two sets of variables, for example to match features to a correct grid reference.

Give map evidence questions

You should look for evidence on the map and make clear statements to support your answer.

Advantages and/or disadvantages questions

You should select relevant advantages or disadvantages of a proposed development, for example the location of a new shopping centre, and demonstrate your understanding of the significance of the proposal.

Some tips for revising

- To be best prepared for the examination, organise your notes into sections. Try to work out a schedule for studying with a programme which includes the sections of the syllabus you intend to study.
- Organise your notes into checklists and revision cards.
- Try to avoid leaving your studying to a day or two before the exam. Also try to avoid cramming your studies into the night before the examination, and especially avoid staying up late to study.
- One useful technique when revising is to use summary note cards on individual topics.
- Make use of past paper questions to test your knowledge or skills. Go over your answers and give yourself a mark for every correct point you make when comparing your answer with your notes.
- If you work with a classmate, try to mark each other's practice answers.
- Practise your diagram-drawing skills and your writing skills. Ensure that your answers are clearly worded. Try to develop the points that you make in your answers.

Some tips for the exam

- Do not write lists, even if you are running out of time. You will lose marks. If the question asks for an opinion based on a choice, for example on the suitability of a particular site or area for a development, do not be afraid to refer to negative points such as why the alternatives are not as good. You will get credit for this.
- Make sure you have a copy of the examination timetable and have planned a schedule for studying.
- Arrive at the examination in plenty of time with the appropriate equipment – pen, pencil, rubber and ruler.
- Carefully read the instructions on the paper and at the beginning of each part of the question.
- Answer all of the compulsory questions in each paper you sit.
- Use the number of marks as a guide to the length of your answer.
- Try to include examples in your answer wherever possible. If asked for diagrams, draw clear, labelled diagrams.
- Read the question instructions very carefully. If the question asks you to "describe", make sure that this is what you do.

- If you are asked to "explain", you must use phrases such as "due to", "this happens because" and "this is a result of". If you describe rather than explain, you will lose most of the marks for that question.
- If you finish early, do not leave the exam. Use the remaining time to check your answers and go over any questions which you have partially answered, especially Ordnance Survey map questions.
- Practise drawing diagrams which may be included in your answers, for example corries or pyramidal peaks.
- Make sure that you have read the instructions on the question carefully and that you have avoided needless errors. For example, answering the wrong sections or failing to explain when asked to, or perhaps omitting to refer to a named area or case study.
- One technique which you might find helpful when answering 5 or 6 mark questions, is to "brainstorm" possible points for your answer. You can write these down in a list at the start of your answer. As you go through your answer, you can double-check with your list to ensure that you have put as much into your answer as you can. This stops you from coming out of the exam and being annoyed that you forgot to mention an important point.

Common errors

Markers of the external examination often remark on errors which occur frequently in candidates' answers. These include the following:

Lack of sufficient detail

- Many candidates fail to provide sufficient detail in answers, often by omitting reference to specific examples, or not elaborating or developing points made in their answer. As noted above, a good guide to the amount of detail required is the number of marks given for the question. If, for example, the total marks offered is 6, then you should make at least six valid points.

Listing

- If you write a simple list of points rather than fuller statements in your answer, you will automatically lose marks. For example, in a 4 mark question, you will obtain only 1 mark for a list.
- The same rule applies to a simple list of bullet points. However, if you couple bullet points with some detailed explanation, you could achieve full marks.

Irrelevant answers

- You must read the question instructions carefully so as to avoid giving answers which are irrelevant to the question. For example, if you are asked to "explain" and you simply "describe", you will lose marks. If you are asked for a named example and you do not provide one, you will forfeit marks.

Repetition

- You should be careful not to repeat points already made in your answer. These will not gain any further marks. You may feel that you have written a long answer, but it may contain the same basic information repeated again and again. Unfortunately, these repeated statements will be ignored by the marker.

Good luck!

Remember that the rewards for passing National 5 Geography are well worth it! Your pass will help you to get the future you want for yourself. In the exam, be confident in your own ability. If you're not sure how to answer a question, trust your instincts and just give it a go anyway. Keep calm and don't panic! GOOD LUCK!

Study Skills – what you need to know to pass exams!

Pause for thought

Many students might skip quickly through a page like this. After all, we all know how to revise. Do you really though?

Think about this:

"IF YOU ALWAYS DO WHAT YOU ALWAYS DO, YOU WILL ALWAYS GET WHAT YOU HAVE ALWAYS GOT."

Do you like the grades you get? Do you want to do better? If you get full marks in your assessment, then that's great! Change nothing! This section is just to help you get that little bit better than you already are.

There are two main parts to the advice on offer here. The first part highlights fairly obvious things but which are also very important. The second part makes suggestions about revision that you might not have thought about but which WILL help you.

Part 1

DOH! It's so obvious but …

Start revising in good time

Don't leave it until the last minute – this will make you panic. Make a revision timetable that sets out work time AND play time.

Sleep and eat!

Obvious really, and very helpful. Avoid arguments or stressful things too – even games that wind you up. You need to be fit, awake and focused!

Know your place!

Make sure you know exactly **WHEN and WHERE** your exams are.

Know your enemy!

Make sure you know what to expect in the exam.

How is the paper structured?

How much time is there for each question?

What types of question are involved?

Which topics seem to come up time and time again?

Which topics are your strongest and which are your weakest?

Are all topics compulsory or are there choices?

Learn by DOING!

There is no substitute for past papers and practice papers – they are simply essential! Tackling this collection of papers and answers is exactly the right thing to be doing as your exams approach.

Part 2

People learn in different ways. Some like low light, some bright. Some like early morning, some like evening / night. Some prefer warm, some prefer cold. But everyone uses their BRAIN and the brain works when it is active. Passive learning – sitting gazing at notes – is the most INEFFICIENT way to learn anything. Below you will find tips and ideas for making your revision more effective and maybe even more enjoyable. What follows gets your brain active, and active learning works!

Activity 1 – Stop and review

Step 1

When you have done no more than 5 minutes of revision reading STOP!

Step 2

Write a heading in your own words which sums up the topic you have been revising.

Step 3

Write a summary of what you have revised in no more than two sentences. Don't fool yourself by saying, "I know it, but I cannot put it into words". That just means you don't know it well enough. If you cannot write your summary, revise that section again, knowing that you must write a summary at the end of it. Many of you will have notebooks full of blue/black ink writing. Many of the pages will not be especially attractive or memorable so try to liven them up a bit with colour as you are reviewing and rewriting. **This is a great memory aid, and memory is the most important thing.**

Activity 2 – Use technology!

Why should everything be written down? Have you thought about "mental" maps, diagrams, cartoons and colour to help you learn? And rather than write down notes, why not record your revision material?

What about having a text message revision session with friends? Keep in touch with them to find out how and what they are revising and share ideas and questions.

Why not make a video diary where you tell the camera what you are doing, what you think you have learned and what you still have to do? No one has to see or hear it, but the process of having to organise your thoughts in a formal way to explain something is a very important learning practice.

Be sure to make use of electronic files. You could begin to summarise your class notes. Your typing might be slow, but it will get faster and the typed notes will be easier to read than the scribbles in your class notes. Try to add different fonts and colours to make your work stand out. You can easily Google relevant pictures, cartoons and diagrams which you can copy and paste to make your work more attractive and **MEMORABLE**.

Activity 3 – This is it. Do this and you will know lots!

Step 1

In this task you must be very honest with yourself! Find the SQA syllabus for your subject (www.sqa.org.uk). Look at how it is broken down into main topics called MANDATORY knowledge. That means stuff you MUST know.

Step 2

BEFORE you do ANY revision on this topic, write a list of everything that you already know about the subject. It might be quite a long list but you only need to write it once. It shows you all the information that is already in your long-term memory so you know what parts you do not need to revise!

Step 3

Pick a chapter or section from your book or revision notes. Choose a fairly large section or a whole chapter to get the most out of this activity.

With a buddy, use Skype, Facetime, Twitter or any other communication you have, to play the game "If this is the answer, what is the question?". For example, if you are revising Geography and the answer you provide is "meander", your buddy would have to make up a question like "What is the word that describes a feature of a river where it flows slowly and bends often from side to side?".

Make up 10 "answers" based on the content of the chapter or section you are using. Give this to your buddy to solve while you solve theirs.

Step 4

Construct a wordsearch of at least 10 × 10 squares. You can make it as big as you like but keep it realistic. Work together with a group of friends. Many apps allow you to make wordsearch puzzles online. The words and phrases can go in any direction and phrases can be split. Your puzzle must only contain facts linked to the topic you are revising. Your task is to find 10 bits of information to hide in your puzzle, but you must not repeat information that you used in Step 3. DO NOT show where the words are. Fill up empty squares with random letters. Remember to keep a note of where your answers are hidden but do not show your friends. When you have a complete puzzle, exchange it with a friend to solve each other's puzzle.

Step 5

Now make up 10 questions (not "answers" this time) based on the same chapter used in the previous two tasks. Again, you must find NEW information that you have not yet used. Now it's getting hard to find that new information! Again, give your questions to a friend to answer.

Step 6

As you have been doing the puzzles, your brain has been actively searching for new information. Now write a NEW LIST that contains only the new information you have discovered when doing the puzzles. Your new list is the one to look at repeatedly for short bursts over the next few days. Try to remember more and more of it without looking at it. After a few days, you should be able to add words from your second list to your first list as you increase the information in your long-term memory.

FINALLY! Be inspired...

Make a list of different revision ideas and beside each one write **THINGS I HAVE** tried, **THINGS I WILL** try and **THINGS I MIGHT** try. Don't be scared of trying something new.

And remember – "FAIL TO PREPARE AND PREPARE TO FAIL!"

NATIONAL 5

2015

National Qualifications 2015

X733/75/11

Geography

THURSDAY, 21 MAY

9:00 AM – 10:45 AM

Total marks — 60

SECTION 1 — PHYSICAL ENVIRONMENTS — 20 marks

Attempt EITHER question 1 **OR** question 2. **ALSO** attempt questions 3, 4 and 5.

SECTION 2 — HUMAN ENVIRONMENTS — 20 marks

Attempt questions 6, 7 and 8

SECTION 3 — GLOBAL ISSUES — 20 marks

Attempt any TWO of the following

Question   9 — Climate Change
Question 10 — Impact of Human Activity on the Natural Environment
Question 11 — Environmental Hazards
Question 12 — Trade and Globalisation
Question 13 — Tourism
Question 14 — Health

Credit will always be given for appropriately labelled sketch maps and diagrams.

Write your answers clearly in the answer booklet provided. In the answer booklet you must clearly identify the question number you are attempting.

Use **blue** or **black** ink.

Before leaving the examination room you must give your answer booklet to the Invigilator; if you do not, you may lose all the marks for this paper.

MARKS

SECTION 1 — PHYSICAL ENVIRONMENTS — 20 marks
Attempt EITHER Question 1 or Question 2
AND Questions 3, 4 and 5

Question 1 — Coastal Landscapes

Study the Ordnance Survey Map Extract (Item A) of the Salcombe area.

(a) Match these grid references with the correct coastal features
Grid references: **766356, 674398, 690382**
Choose from features: cliff; headland; bay; stack.

3

(b) **Explain** the formation of **one** of the coastal features listed in part (a).
You may use a diagram(s) in your answer.

4

NOW ANSWER QUESTIONS 3, 4 AND 5

DO NOT ANSWER THIS QUESTION IF YOU HAVE ALREADY ANSWERED QUESTION 1

Question 2 — Rivers and Valleys

Study the Ordnance Survey Map Extract (Item A) of the Salcombe area.

(a) Match these grid references with the correct river features
Grid references: **708473, 713410, 684466**
Choose from features: levée; meander; v-shaped valley; waterfall.

3

(b) **Explain** the formation of **one** of the river features listed in part (a).
You may use a diagram(s) in your answer.

4

NOW ANSWER QUESTIONS 3, 4 AND 5

MARKS

Question 3

Diagram Q3: Quote from a Local Landowner

"This area has the potential for a variety of different land uses, including farming, forestry, recreation/tourism, water storage/supply, industry and renewable energy."

Study Diagram Q3 and the Ordnance Survey Map Extract (Item A) of the Salcombe area.

Choose **two** different land uses shown in Diagram Q3.

Using map evidence, **explain** how the area shown on the map extract is suitable for your chosen land uses.

5

Question 4

Diagram Q4: Selected Land Uses

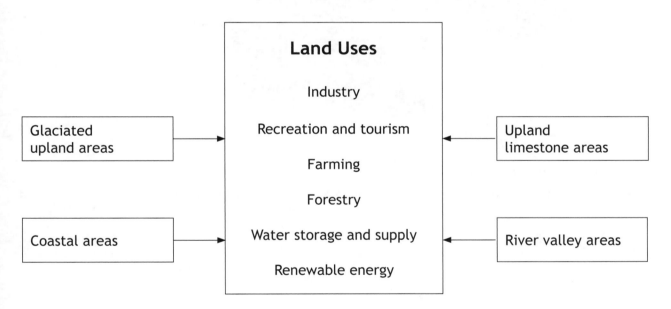

Look at Diagram Q4 above.

For a named area you have studied, **explain**, **in detail**, ways in which **two** different land uses may be in conflict with each other.

4

[Turn over

MARKS

Question 5

Diagram Q5: Average UK Temperatures in July

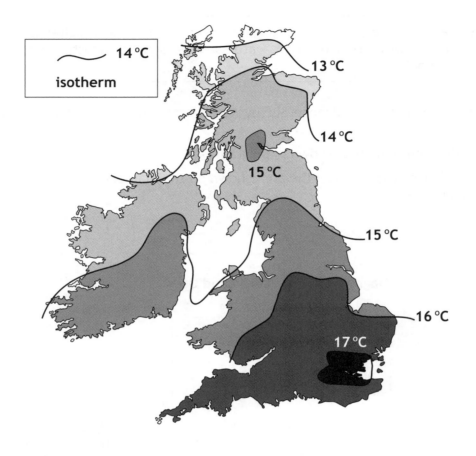

Look at Diagram Q5.

Explain the factors which cause differences in average UK temperatures.

4

MARKS

SECTION 2 — HUMAN ENVIRONMENTS — 20 marks
Attempt Questions 6, 7 and 8

Question 6

Diagram Q6

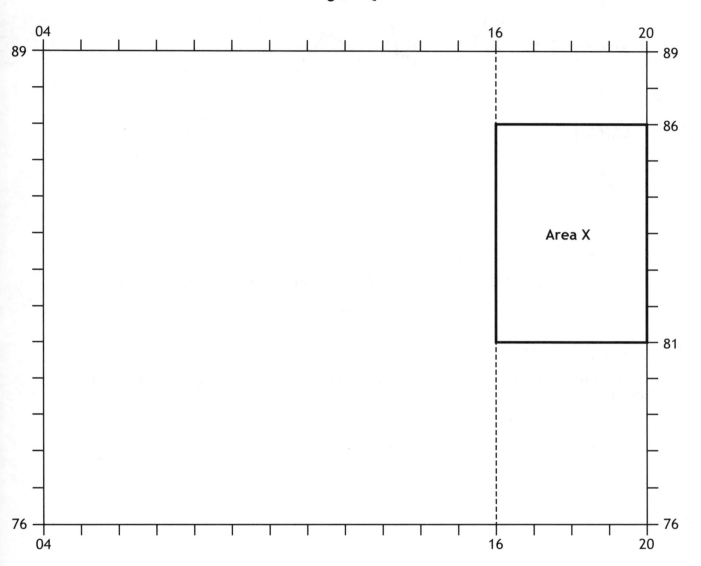

Study the Ordnance Survey Map Extract (Item B) of the Birmingham area and Diagram Q6 above.

(a) Give map evidence to show that part of the Central Business District (CBD) of Birmingham is found in grid square 0786.

3

(b) Find Area X on Diagram Q6 and the map extract (Item B).

Birmingham Airport, a golf course, a business park and a housing area are found in Area X on the rural/urban fringe of Birmingham. Using map evidence **explain** why such developments are found there.

5

Question 7

Diagram Q7: Births in Scotland 1901–2011

Look at Diagram Q7

Give reasons why the birth rate has decreased in developed countries such as Scotland.

6

MARKS

Question 8

Diagram Q8: Shanty Town Improvements in Brazil

Look at Diagram Q8

For a named city in the developing world **describe**, **in detail**, measures taken to improve conditions in shanty towns.

6

[Turn over

SECTION 3 — GLOBAL ISSUES — 20 marks

Attempt any TWO questions

MARKS

Question 9 — Climate Change

Diagram Q9: Area of Arctic Sea Ice (1979–2013)

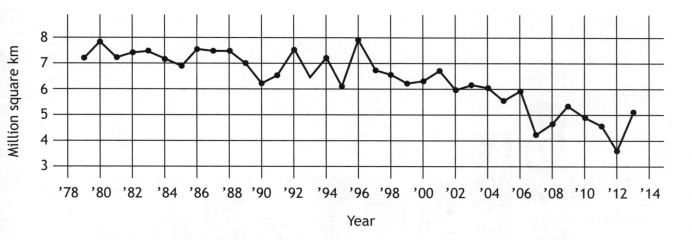

Study Diagram Q9

(a) **Describe**, **in detail**, the changes in the area of Arctic Sea ice.   4

(b) Melting sea ice is one effect of climate change.

   Explain some other effects of climate change.   6

[Turn over

MARKS

Question 10 — Impact of Human Activity on the Natural Environment

Diagram Q10A: Deforestation in Peru 2004–2012

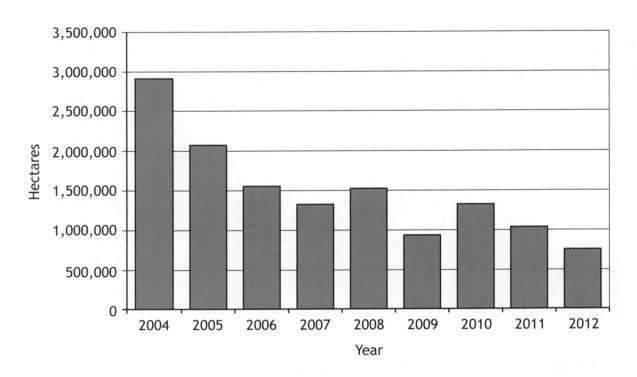

(a) Study Diagram Q10A.

Describe, in detail, the changes in deforestation in Peru from 2004 to 2012.

4

Diagram Q10B: Human Activity in the Tundra and Equatorial regions.

Oil pipeline in the Tundra

Cattle ranching in the Rainforest

(b) Look at Diagram Q10B.

For a named area you have studied, **explain** the impact of recent human activity on people and the environment.

6

MARKS

Question 11 — Environmental Hazards

Diagram Q11A: Earthquake Threatened Cities

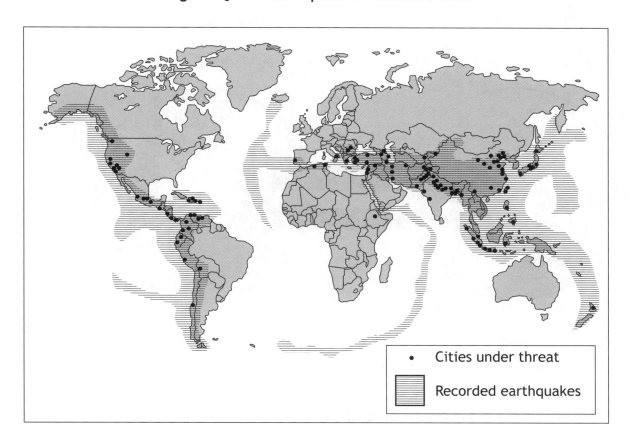

- • Cities under threat
- ▤ Recorded earthquakes

(a) Study Diagram Q11A.

Describe, **in detail**, the distribution of cities most threatened by earthquakes.   **4**

Diagram Q11B

(b) Look at Diagram Q11B.

Explain, **in detail**, the strategies used to reduce the impact of an earthquake.

You must refer to named examples you have studied in your answer.   **6**

Question 12 — Trade and Globalisation   MARKS

Diagram Q12A: Pattern of World Trade

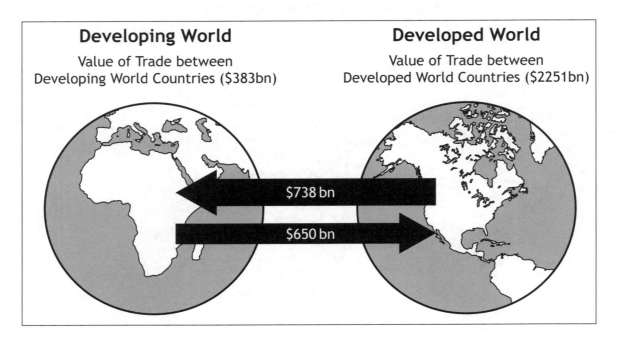

Developing World	Developed World
Value of Trade between Developing World Countries ($383bn)	Value of Trade between Developed World Countries ($2251bn)

$738 bn

$650 bn

(a) Study Diagram Q12A.

Describe, **in detail**, the pattern of world trade.   4

Diagram Q12B: Trade Between Africa and the European Union (EU)

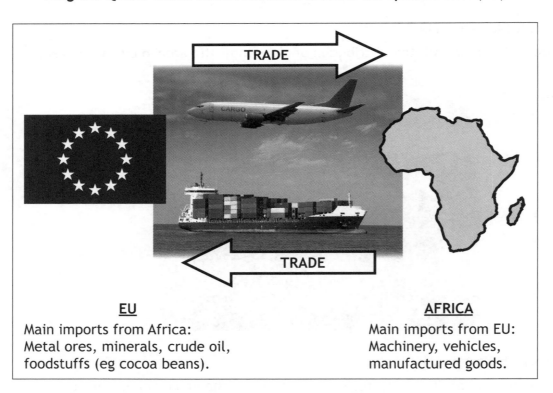

TRADE

TRADE

EU	**AFRICA**
Main imports from Africa: Metal ores, minerals, crude oil, foodstuffs (eg cocoa beans).	Main imports from EU: Machinery, vehicles, manufactured goods.

(b) Look at Diagram Q12B.

Referring to example(s) you have studied, **describe** the impact of world trade on people and the environment.   6

MARKS

Question 13 — Tourism

Diagram Q13A: Global Visitor Numbers: 1995–2013

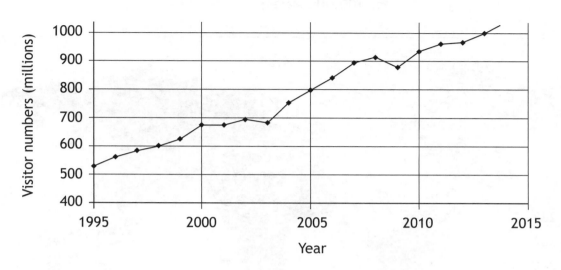

(a) Study Diagram Q13A.

Describe, in detail, the changes in global visitor numbers since 1995.　　　4

Diagram Q13B: Quote from Rainforest Community Leader

"ECO-TOURISM has helped us to support environmental protection and improve the well-being of our people all year round."

(b) Look at Diagram Q13B.

For a named tourist area you have studied, describe, in detail the impact of eco-tourism on people and the environment.　　　6

[Turn over for Question 14 on *Page fourteen*

MARKS

Question 14 — Health

**Diagram Q14A: Worldwide Male Deaths from Heart Disease in 2011
(per 100,000 Males)**

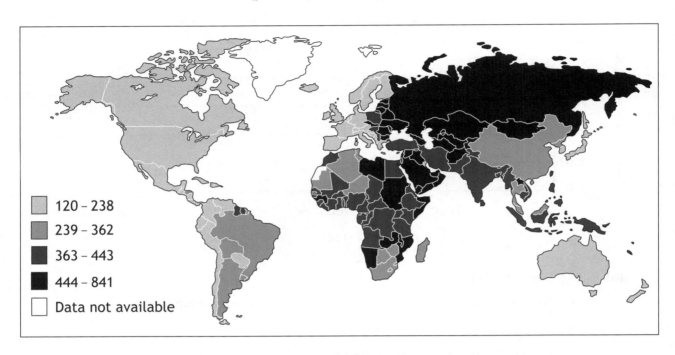

120 – 238

239 – 362

363 – 443

444 – 841

Data not available

(a) Study Diagram Q14A.

Describe, **in detail**, the distribution of male deaths from heart disease.   4

Diagram Q14B: Selected Developing World Diseases

- **Malaria**
- **Cholera**
- **Kwashiorkor**
- **Pneumonia**

(b) Choose **one** disease from Diagram Q14B above.

For the disease you have chosen, **explain** the methods used to control it.   6

[END OF QUESTION PAPER]

National Qualifications 2015

X733/75/21

Geography
Ordnance Survey Map
Item A

THURSDAY, 21 MAY

9:00 AM – 10:45 AM

The colours used in the printing of these map extracts are indicated in the four little boxes at the top of the map extract. Each box should contain a colour; if any does not, the map is incomplete and should be returned to the Invigilator.

Extract No 2143/202

1:50 000 Scale
Landranger Series

Four colours should appear above; if not then please return to the invigilator.

ROADS AND PATHS

Not necessarily rights of way

Service area, Junction number, Elevated, M1	Motorway (dual carriageway)
Unfenced A 470, Dual carriageway	Primary Route (recommended through route)
A 493, Footbridge	Main road
	Road under construction
B 4518	Secondary road
A 855, Bridge, B 885	Narrow road with passing places
	Road generally more than 4m wide
	Road generally less than 4m wide
	Path / Other road, drive or track
	Gradient: steeper than 20% (1 in 5), 14% to 20% (1 in 7 to 1 in 5)
	Gates, Road tunnel
Ferry P, Ferry V	Ferry (passenger), Ferry (vehicle)

RAILWAYS

	Track multiple or single		Bridges, footbridge
	Track under construction	LC	Level crossing
	Siding		Viaduct, embankment
	Tunnel, cuttings	a	Station, (a) principal
	Light rapid transit system, narrow gauge or tramway		Light rapid transit system station

WATER FEATURES

Marsh or salting, Towpath, Lock, Slopes, Cliff, Shingle, Aqueduct, Canal, Ford, Beacon, Flat rock, Lighthouse (in use), Lake, Weir, Footbridge, Bridge, Normal tidal limit, Sand Dunes, Lighthouse (disused), Low water mark, Mud, High water mark, Canal (dry)

HEIGHTS

1 metre = 3·2808 feet

Contours are at 10 metres vertical interval — 50

·144 Heights are to the nearest metre above mean sea level

Where two heights are shown the first height is to the base of the triangulation pillar and the second (in brackets) to the highest natural point of the hill

ROCK FEATURES

Outcrop, Cliff, Scree

PUBLIC RIGHTS OF WAY

·········	Footpath
————————	Bridleway
———————	Restricted byway
-+-+-+-+-	Byway open to all traffic

The symbols show the defined route so far as the scale of mapping will allow.

The representation on this map of any other road, track or path is no evidence of the existence of a right of way. Not shown on maps of Scotland

Danger Area Firing and Test Ranges in the area. Danger! Observe warning notices.

OTHER PUBLIC ACCESS

• • • •	Other route with public access (not normally shown in urban areas). Alignments are based on the best information available. These routes are not shown on maps of Scotland.
● ●	On-road cycle route
○ ○	Traffic-free cycle route
4	National Cycle Network number
8	Regional Cycle Network number
◆ ◆	National Trail, European Long Distance Path, Long Distance Route, selected Recreational Routes

BOUNDARIES

—+—+—+—	National
—+—+—+—	District
—·—·—·—	County, Unitary Authority, Metropolitan District or London Borough
	National Park

ANTIQUITIES

+	Site of antiquity
✕	Battlefield (with date)
☆ ····	Visible earthwork
VILLA	Roman
Castle	Non-Roman

TOURIST INFORMATION

⚑ ⚏ ⚏	Camp site / caravan site
❀	Garden
⛳	Golf course or links
i i	Information centre (all year / seasonal)
⚘	Nature reserve
P P&R P&R	Parking, Park and ride (all year / seasonal)
⚒	Picnic site
⚽	Recreation / leisure / sports centre
▨	Selected places of tourist interest
☎ ☎	Telephone, public / roadside assistance
☀	Viewpoint
V	Visitor centre
⚑	Walks / Trails
◉	World Heritage site or area
▲	Youth hostel

LAND FEATURES

⋏ ⋏	Electricity transmission line (pylons shown at standard spacing)
> – > – – >	Pipe line (arrow indicates direction of flow)
ruin	Buildings
	Important building (selected)
	Bus or coach station
	Current or former place of worship { with tower, with spire, minaret or dome }
+	Place of worship
⊘	Glass structure
(H)	Heliport
△	Triangulation pillar
Ŧ	Mast
ⵣ	Wind pump, wind turbine
ⵣ	Windmill with or without sails
+	Graticule intersection at 5' intervals
	Cutting, embankment
	Landfill site or slag/spoil heap
	Coniferous wood
	Non-coniferous wood
	Mixed wood
	Orchard
	Park or ornamental ground
	Forestry Commission land
	National Trust (always open / limited access, observe local signs)
	National Trust for Scotland (always open / limited access, observe local signs)

ABBREVIATIONS

Br	Bridge	MS	Milestone
Cemy	Cemetery	Mus	Museum
CG	Cattle grid	P	Post office
CH	Clubhouse	PC	Public convenience (in rural areas)
Fm	Farm	PH	Public house
Ho	House	Sch	School
MP	Milepost	TH	Town Hall, Guildhall or equivalent

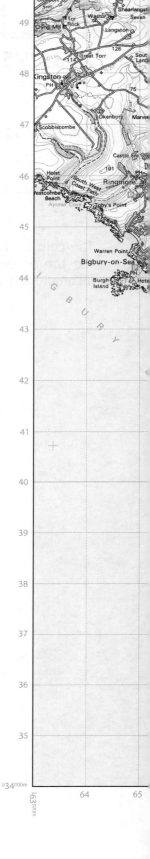

Scale 1:50 000

[BLANK PAGE]

DO NOT WRITE ON THIS PAGE

National Qualifications 2015

X733/75/31

Geography
Ordnance Survey Map
Item B

THURSDAY, 21 MAY

9:00 AM – 10:45 AM

The colours used in the printing of these map extracts are indicated in the four little boxes at the top of the map extract. Each box should contain a colour; if any does not, the map is incomplete and should be returned to the Invigilator.

Extract No 2142/139

1:50 000 Scale
Landranger Series

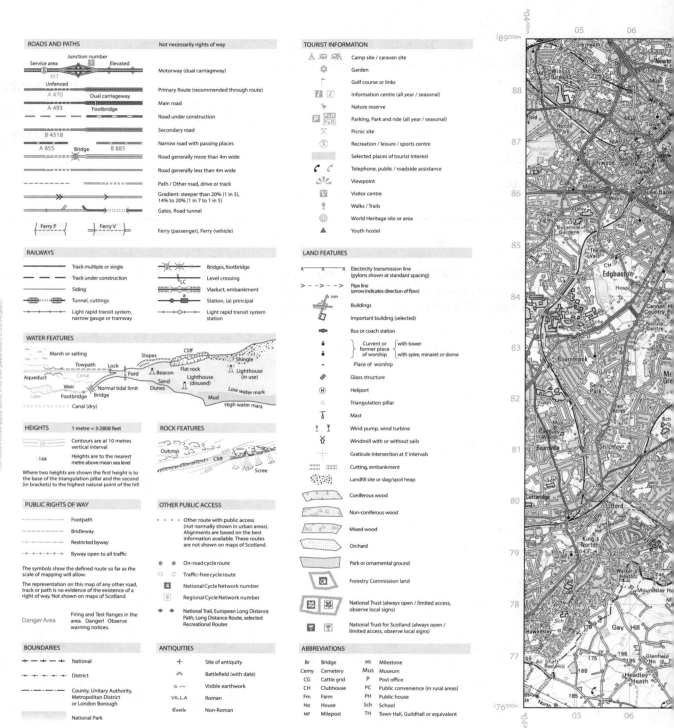

ROADS AND PATHS

Not necessarily rights of way

Service area | Junction number | Elevated
M1
Motorway (dual carriageway)

Unfenced
A 470 | Dual carriageway | Primary Route (recommended through route)
A 493 | Footbridge | Main road
Road under construction
Secondary road
B 4518
A 855 | Bridge | B 885 | Narrow road with passing places
Road generally more than 4m wide
Road generally less than 4m wide
Path / Other road, drive or track
Gradient: steeper than 20% (1 in 5), 14% to 20% (1 in 7 to 1 in 5)
Gates, Road tunnel

Ferry P | Ferry V | Ferry (passenger), Ferry (vehicle)

RAILWAYS

Track multiple or single | Bridges, footbridge
Track under construction | Level crossing | LC
Siding | Viaduct, embankment
Tunnel, cuttings | Station, (a) principal
Light rapid transit system, narrow gauge or tramway | Light rapid transit system station

WATER FEATURES

Marsh or salting
Towpath | Slopes | Cliff | Shingle
Lock | Flat rock
Aqueduct | Canal | Ford | Beacon | Lighthouse (in use)
Lake | Weir | Sand | Lighthouse (disused)
Footbridge | Bridge | Normal tidal limit | Dunes | Mud | Low water mark
Canal (dry) | High water mark

HEIGHTS

1 metre = 3·2808 feet

50 | Contours are at 10 metres vertical interval
·144 | Heights are to the nearest metre above mean sea level

Where two heights are shown the first height is to the base of the triangulation pillar and the second (in brackets) to the highest natural point of the hill

ROCK FEATURES

Outcrop
Cliff
Scree

PUBLIC RIGHTS OF WAY

Footpath
Bridleway
Restricted byway
Byway open to all traffic

The symbols show the defined route so far as the scale of mapping will allow.

The representation on this map of any other road, track or path is no evidence of the existence of a right of way. Not shown on maps of Scotland

Danger Area | Firing and Test Ranges in the area. Danger! Observe warning notices.

OTHER PUBLIC ACCESS

Other route with public access (not normally shown in urban areas). Alignments are based on the best information available. These routes are not shown on maps of Scotland.

On-road cycle route
Traffic-free cycle route
4 | National Cycle Network number
8 | Regional Cycle Network number
National Trail, European Long Distance Path, Long Distance Route, selected Recreational Routes

BOUNDARIES

National
District
County, Unitary Authority, Metropolitan District or London Borough
National Park

ANTIQUITIES

+ | Site of antiquity
✕ | Battlefield (with date)
☆ ···· | Visible earthwork
VILLA | Roman
Castle | Non-Roman

TOURIST INFORMATION

⛺ | Camp site / caravan site
🌼 | Garden
Golf course or links
i | Information centre (all year / seasonal)
Nature reserve
P P&R | Parking, Park and ride (all year / seasonal)
✗ | Picnic site
Recreation / leisure / sports centre
Selected places of tourist interest
☎ | Telephone, public / roadside assistance
Viewpoint
V | Visitor centre
! | Walks / Trails
World Heritage site or area
▲ | Youth hostel

LAND FEATURES

× —— × | Electricity transmission line (pylons shown at standard spacing)
> – – > – > | Pipe line (arrow indicates direction of flow)
ruin | Buildings
Important building (selected)
Bus or coach station
Current or former place of worship | with tower | with spire, minaret or dome
+ | Place of worship
Glass structure
H | Heliport
△ | Triangulation pillar
Ŏ | Mast
Ψ Ψ | Wind pump, wind turbine
Ŭ | Windmill with or without sails
Graticule intersection at 5' intervals
Cutting, embankment
Landfill site or slag/spoil heap
Coniferous wood
Non-coniferous wood
Mixed wood
Orchard
Park or ornamental ground
Forestry Commission land
National Trust (always open / limited access, observe local signs)
National Trust for Scotland (always open / limited access, observe local signs)

ABBREVIATIONS

Br	Bridge	MS	Milestone
Cemy	Cemetery	Mus	Museum
CG	Cattle grid	P	Post office
CH	Clubhouse	PC	Public convenience (in rural areas)
Fm	Farm	PH	Public house
Ho	House	Sch	School
MP	Milepost	TH	Town Hall, Guildhall or equivalent

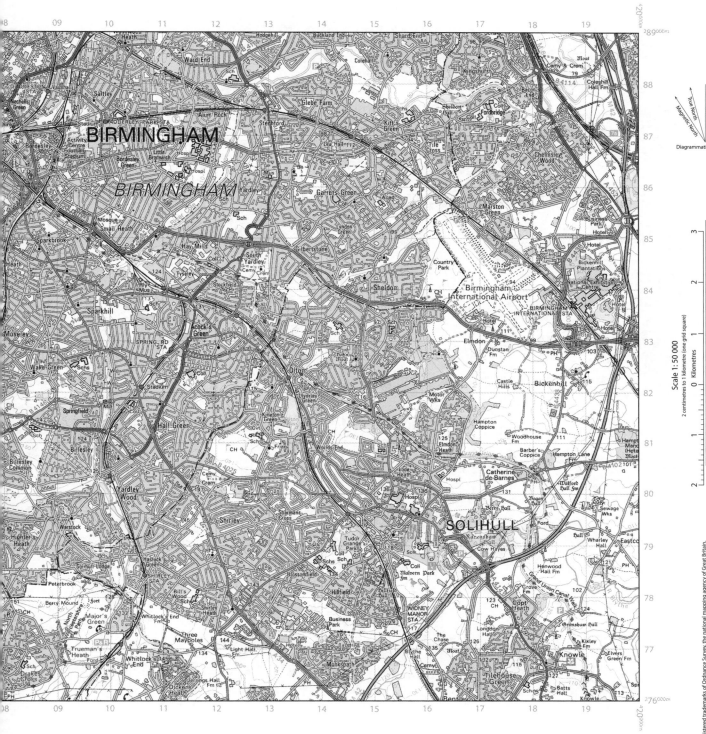

Scale 1:50 000

[BLANK PAGE]

DO NOT WRITE ON THIS PAGE

NATIONAL 5

2017

National Qualifications 2017

X733/75/11   Geography

FRIDAY, 26 MAY

1:00 PM – 2:45 PM

Total marks — 60

SECTION 1 — PHYSICAL ENVIRONMENTS — 20 marks

Attempt EITHER Question 1 OR Question 2. ALSO attempt Questions 3 and 4.

SECTION 2 — HUMAN ENVIRONMENTS — 20 marks

Attempt Questions 5, 6, 7 and 8.

SECTION 3 — GLOBAL ISSUES — 20 marks

Attempt any TWO of the following.

Question   9 — Climate Change

Question 10 — Impact of Human Activity on the Natural Environment

Question 11 — Environmental Hazards

Question 12 — Trade and Globalisation

Question 13 — Tourism

Question 14 — Health

Credit will be given for appropriately labelled sketch maps and diagrams.

Write your answers clearly in the answer booklet provided. In the answer booklet you must clearly identify the question number you are attempting.

Use blue or black ink.

Before leaving the examination room you must give your answer booklet to the Invigilator; if you do not, you may lose all the marks for this paper.

MARKS

SECTION 1 — PHYSICAL ENVIRONMENTS — 20 marks
Attempt EITHER Question 1 OR Question 2
ALSO attempt Questions 3 and 4

Question 1: Glaciated Upland

Diagram Q1: Glacial contour patterns

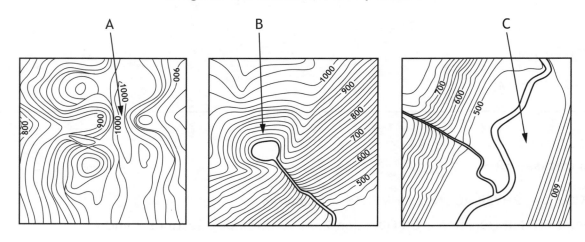

(a) Match the letters on Diagram Q1 with the correct glacial features below.

Choose from:

| U-shaped valley | corrie | pyramidal peak | arête | 3 |

(b) **Explain** the processes involved in the formation of a U-shaped valley.
You may use a diagram(s) in your answer.   4

(c) **Explain** different ways in which people use glaciated landscapes.   4

NOW ATTEMPT QUESTIONS 3 AND 4

MARKS

DO NOT ATTEMPT THIS QUESTION IF YOU HAVE ALREADY ANSWERED QUESTION 1

Question 2: Upland Limestone

Diagram Q2: Upland limestone landscape

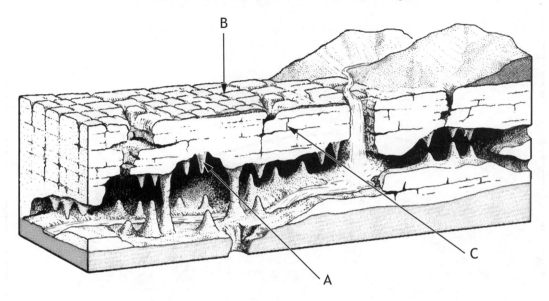

(a) Match the letters on Diagram Q2 with the correct limestone features below.

Choose from:

| stalactite | stalagmite | clint | grike | joint | bedding plane | 3 |

(b) **Explain** the formation of a limestone pavement.

You may use a diagram(s) in your answer.   4

(c) **Explain** different ways in which people use limestone landscapes.   4

NOW ATTEMPT QUESTIONS 3 AND 4

[Turn over

MARKS

Question 3

Diagram Q3: Synoptic chart for 12.00 on 28th December 2014

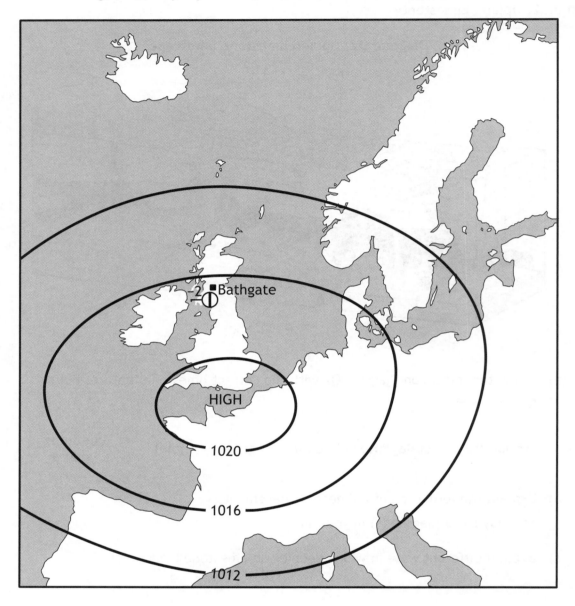

Study Diagram Q3.

Give reasons for the weather conditions at Bathgate on 28th December 2014.   5

Question 4

Anticyclones bring different weather conditions throughout the year.

Describe the benefits **and** problems of an anticyclone in **summer**.   4

MARKS

SECTION 2 — HUMAN ENVIRONMENTS — 20 marks
Attempt Questions 5, 6, 7 and 8

Question 5

Study the Ordnance Survey map extract (Item A) of the Edinburgh area.

Match the grid references with the correct urban land use zone.

Grid references: **2568, 2573, 2671**

Choose from the urban land use zones below.

- **CBD**
- **new industry**
- **new housing**
- **old housing**   3

Question 6

Study the Ordnance Survey map extract (Item A) of the Edinburgh area.

There is a plan to build new housing in grid square 2667.

Using map evidence, **explain** why this area is suitable for new housing.   5

[Turn over

MARKS

Question 7

Diagram Q7: World Population Density

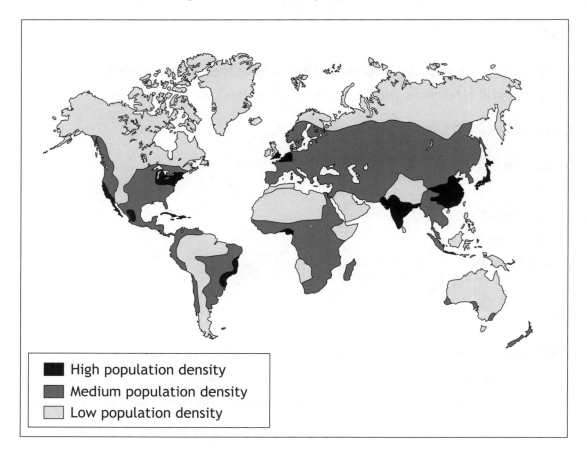

Look at Diagram Q7.

Explain why there are areas of different population density across the world.

Your answer should refer to both physical **and** human factors.

6

MARKS

Question 8

Diagram Q8: Shanty Town

De Visu / Shutterstock.com

Look at Diagram Q8.

Referring to an area you have studied, **describe** different ways shanty towns are being improved.

6

[Turn over

SECTION 3 — GLOBAL ISSUES — 20 marks
Attempt any TWO questions

MARKS

Question 9 — Climate Change

Diagram Q9A: Average Global Temperature Change 1996—2016

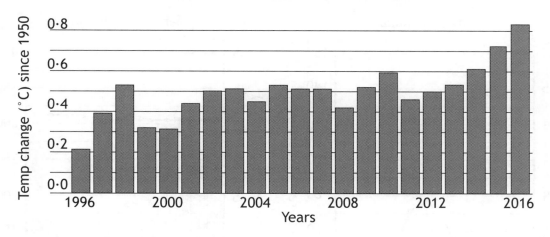

(a) Study Diagram Q9A.

   Describe, in detail, average global temperature change from 1996 to 2016.   4

Diagram Q9B: Newspaper Headline

(b) Look at Diagram Q9B.

   Explain, in detail, strategies used to minimise future climate change.   6

[Turn over

MARKS

Question 10 — Impact of Human Activity on the Natural Environment

Diagram Q10

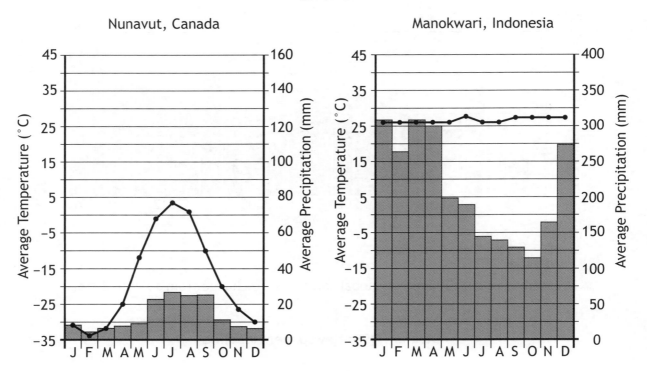

(a) Study Diagram Q10.

Use the information in Diagram Q10 to **describe**, **in detail**, the differences between the two climates shown.   **4**

(b) For a named **tundra** or **equatorial** area which you have studied, **explain** the impact of human activity on people **and** the environment.   **6**

MARKS

Question 11 — Environmental Hazards

Diagram Q11A: World Distribution of Tropical Storms

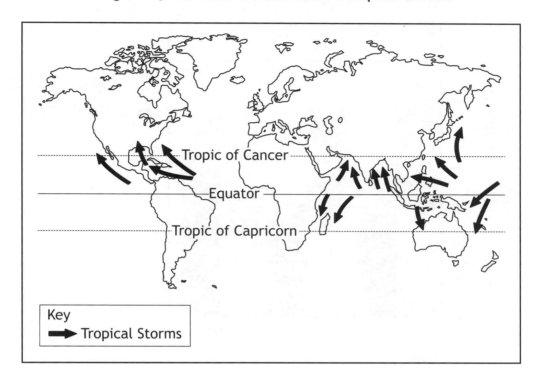

(a) Study Diagram Q11A.

Describe, in detail, the distribution of tropical storms.   4

Diagram Q11B: A Tropical Storm hits coastal town

Fabio Lamanna / Shutterstock.com

(b) Look at Diagram Q11B.

For a tropical storm you have studied, **explain in detail** the impacts of the storm on people **and** the environment.   6

MARKS

Question 12 — Trade and Globalisation

Diagram Q12A: Percentage Share of World Trade 1995–2010

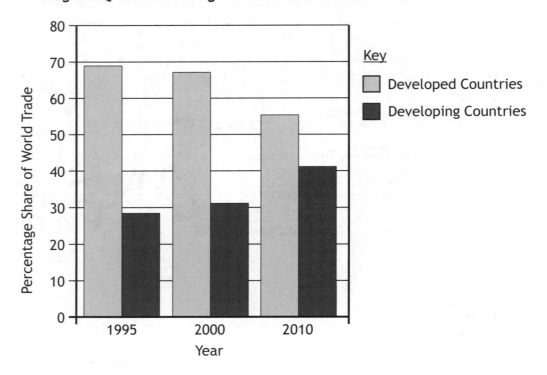

(a) Study Diagram Q12A.

Describe, **in detail**, the changes in percentage share of world trade from 1995–2010.

4

Diagram Q12B: Trade Report

Trade Report
Trade Inequality increases between Developing and Developed countries in 2015.

(b) Look at Diagram Q12B.

Explain, **in detail**, the causes of inequalities in trade between developed and developing countries.

6

Question 13 — Tourism

Diagram Q13A: Origin of Tourists Visiting Scotland (thousands)

COUNTRY	2006	2010	2014
USA	475	275	418
Germany	278	253	343
France	229	196	190
Australia	133	147	158
Netherlands	114	135	149
Canada	161	98	122
Ireland	224	185	113
Spain	142	139	101
Rest of World	976	913	1,106
TOTAL	**2,732**	**2,341**	**2,700**

(a) Study Diagram Q13A.

Describe, in detail, the changes in the number of tourists visiting Scotland from different countries between 2006 and 2014.

4

Diagram Q13B: Quote from a tour operator

"Mass tourism has increased since the 1950s with many locations at home and abroad experiencing a record number of visitors year on year."

(b) Look at Diagram Q13B.

Give reasons for the increase in mass tourism.

6

[Turn over

MARKS

Question 14 — Health

Diagram Q14A: Percentage Change in Death Rates from Malaria, 2000–2013

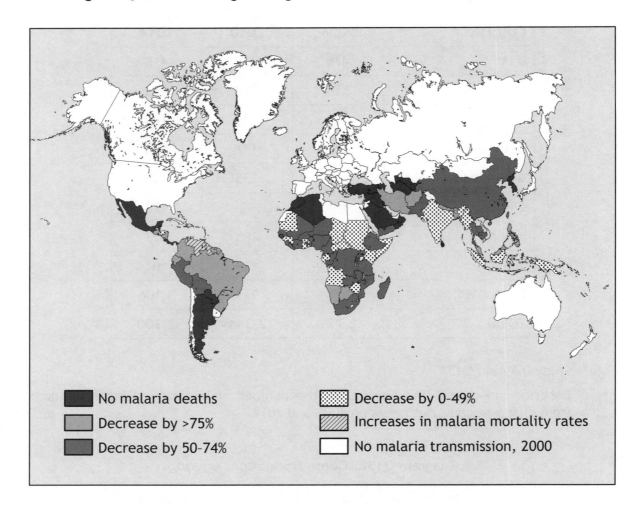

■ No malaria deaths	▦ Decrease by 0–49%
Decrease by >75%	▨ Increases in malaria mortality rates
Decrease by 50–74%	☐ No malaria transmission, 2000

(a) Study Diagram Q14A.

    Describe, in detail, the changes in death rates from malaria.   **4**

MARKS

Question 14 — Health (continued)

Diagram Q14B: Selected Developing World Diseases

cholera
pneumonia
malaria
kwashiorkor

(b) Look at Diagram Q14B.

Choose **one** disease from Diagram Q14B above.

For the disease you have chosen, explain the impact on people **and** the countries affected.

6

[END OF QUESTION PAPER]

MARKS

[BLANK PAGE]

DO NOT WRITE ON THIS PAGE

National
Qualifications
2017

X733/75/21

Geography
Ordnance Survey Map
Item A

FRIDAY, 26 MAY

1:00 PM – 2:45 PM

The colours used in the printing of these map extracts are indicated in the four little boxes at the top of the map extract. Each box should contain a colour; if any does not, the map is incomplete and should be returned to the Invigilator.

Ordnance Survey

1:50 000 Scale
Landranger Series

ROADS AND PATHS

Not necessarily rights of way

Junction number
Service area   Elevated
M1
Unfenced
A 470
A 493
B 4518
A 855   Bridge   B 885

Motorway (dual carriageway)

Primary Route
(A network of recommended through routes which complement the motorway system)

Main road

Road under construction

Secondary road

Narrow road with passing places

Road generally more than 4m wide

Road generally less than 4m wide

Path / Other road, drive or track

Gradient: steeper than 20% (1 in 5), 14% to 20% (1 in 7 to 1 in 5)

Gates, Road tunnel

Ferry P   Ferry V

Ferry (passenger), Ferry (vehicle)

Dual carriageway

Footbridge

RAILWAYS

Track multiple or single
Track under construction
Siding
Tunnel, cuttings
Narrow gauge, tramway or light rail system

Bridges, footbridge
Level crossing
Viaduct, embankment
Station, (a) principal
Light rail station

WATER FEATURES

Marsh or salting
Towpath   Lock
Aqueduct
Weir
Footbridge   Bridge
Lake
Canal (dry)

Slopes   Cliff
Shingle
Flat rock
Ford   Beacon   Sand
Lighthouse
Dunes   (disused)
Low water mark
Mud
High water mark
Lighthouse (in use)
Normal tidal limit
Canal

HEIGHTS

1 metre = 3·2808 feet

Contours are at 10 metres vertical interval

·144   Heights are to the nearest metre above mean sea level

Where two heights are shown, the first is the height of the natural ground in the location of the triangulation pillar, and the second (in brackets) to a separate point which is the natural summit.

ROCK FEATURES

Outcrop
Cliff
Scree

PUBLIC RIGHTS OF WAY

Footpath
Bridleway
Restricted byway (not for use by mechanically propelled vehicles)
Byway open to all traffic

The symbols show the defined route so far as the scale of mapping will allow.

The representation on this map of any other road, track or path is no evidence of the existence of a right of way. Not shown on maps of Scotland

Danger Area   Firing and Test Ranges in the area. Danger! Observe warning notices.

OTHER PUBLIC ACCESS

• • • •   Other route with public access (not normally shown in urban areas). Alignments are based on the best information available. These routes are not shown on maps of Scotland.

On-road cycle route
Traffic-free cycle route
4   National Cycle Network number
8   Regional Cycle Network number
National Trail, Scotland's Great Trails, European Long Distance Path and selected Recreational Routes

BOUNDARIES

National
District
County, Unitary Authority, Metropolitan District or London Borough
National Park

ANTIQUITIES

+   Site of antiquity
⚔   Site of Battle (with date)
☆ ····   Visible earthwork
VILLA   Roman
Castle   Non-Roman

TOURIST INFORMATION

Camp site / caravan site
Garden/aboretum
Golf course or links
Information centre (all year / seasonal)
Nature reserve
Parking, Park and ride (all year / seasonal)
Picnic site
Recreation / leisure / sports centre
Selected places of tourist interest
Phone, public / emergency
Viewpoint
Visitor centre
Walks / Trails
World Heritage site or area
Youth hostel

LAND FEATURES

Electricity transmission line (pylons shown at standard spacing)
Pipe line (arrow indicates direction of flow)
ruin
Buildings
Important building (selected)
Bus or coach station
Current or former place of worship } with tower / with spire, minaret or dome
Place of worship
Glass structure
Heliport
Triangulation pillar
Mast
Wind pump
Wind turbine
Windmill with or without sails
Graticule intersection at 5' intervals
Cutting, embankment
Landfill site or slag/spoil heap
Coniferous wood
Non-coniferous wood
Mixed wood
Orchard
Park or ornamental ground
Forestry Commission land
National Trust (always open / limited access, observe local signs)
Natural Resources Wales
National Trust for Scotland (always open / limited access, observe local signs)

ABBREVIATIONS

Br	Bridge	MS	Milestone
Cemy	Cemetery	Mus	Museum
CG	Cattle grid	P	Post office
CH	Clubhouse	PC	Public convenience (in rural areas)
Fm	Farm	PH	Public house
Hospl	Hospital	Sch	School
Ho	House	TH	Town Hall, Guildhall or equivalent
MP	Milepost	Univ	University

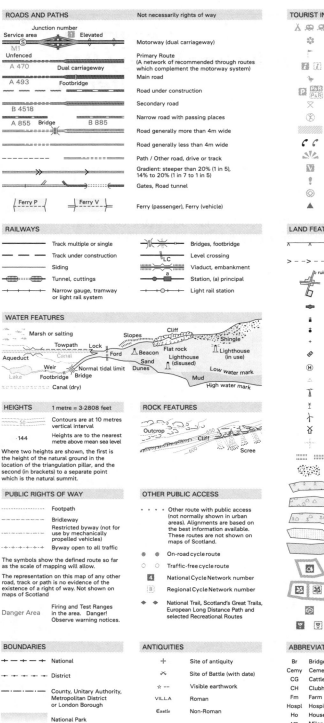

Magnetic North   Grid North   True North

Diagrammatic only

Extract No 2248/66

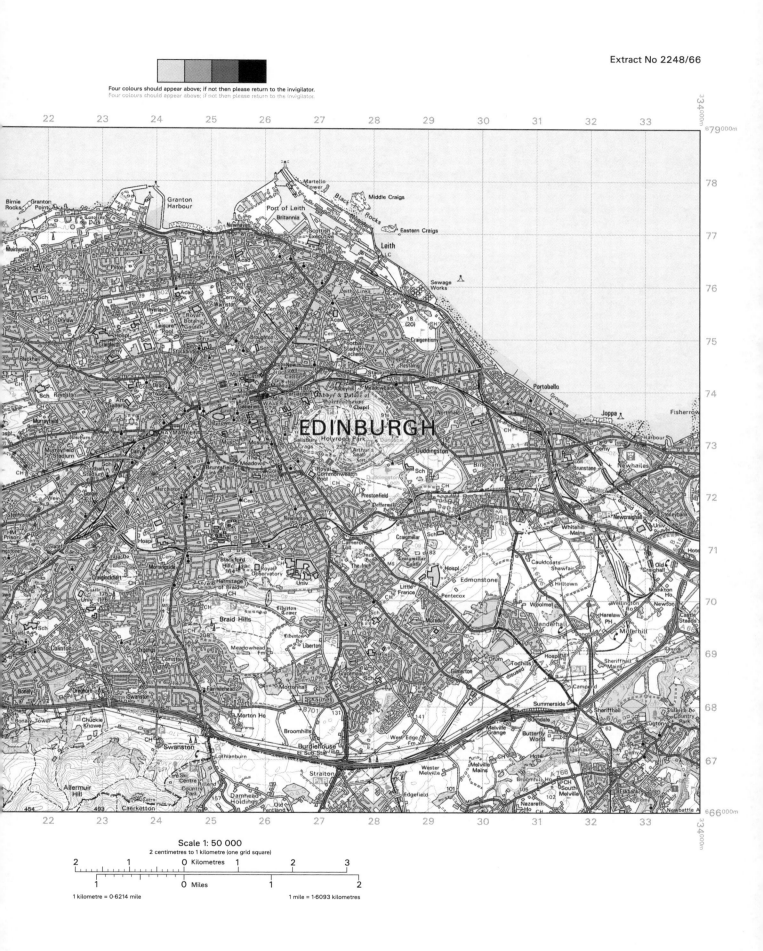

Scale 1: 50 000
2 centimetres to 1 kilometre (one grid square)

Kilometres

Miles

1 kilometre = 0·6214 mile

1 mile = 1·6093 kilometres

[BLANK PAGE]

DO NOT WRITE ON THIS PAGE

NATIONAL 5

2017 Specimen
Question Paper

National Qualifications SPECIMEN ONLY

S833/75/11 **Geography**

Date — Not applicable

Duration — 2 hours 20 minutes

Total marks — 80

SECTION 1 — PHYSICAL ENVIRONMENTS — 30 marks

Attempt **EITHER** question 1 **OR** question 2

THEN attempt questions 3 to 6.

SECTION 2 — HUMAN ENVIRONMENTS — 30 marks

Attempt ALL questions.

SECTION 3 — GLOBAL ISSUES — 20 marks

Attempt any **TWO** of the following.

Question 11 — Climate change

Question 12 — Natural regions

Question 13 — Environmental hazards

Question 14 — Trade and globalisation

Question 15 — Tourism

Question 16 — Health

You will receive credit for appropriately labelled sketch maps and diagrams.

Write your answers clearly in the answer booklet provided. In the answer booklet you must clearly identify the question number you are attempting.

Use **blue** or **black** ink.

Before leaving the examination room you must give your answer booklet to the Invigilator; if you do not, you may lose all the marks for this paper.

MARKS

SECTION 1 — PHYSICAL ENVIRONMENTS — 30 marks
Attempt EITHER question 1 OR question 2
THEN questions 3 to 6

Question 1 — Glaciated landscapes

(a) Study the Ordnance Survey map extract (Item A) of the Brecon Beacons area.

Using grid references, **describe** the evidence shown on the map which suggests that this is an area of **upland glaciated scenery**. **4**

(b) **Explain** the formation of a **U-shaped valley**.

You may use a diagram(s) in your answer. **4**

Now attempt questions 3 to 6

MARKS

Do not attempt question 2 if you have already answered question 1

Question 2 — Rivers and valleys

(a) Study the Ordnance Survey map extract (Item A) of the Brecon Beacons area.

Describe the physical features of the Afon (River) Nedd Fechan **and** its valley between 905175 and 900092. You should use grid references in your answer.   4

(b) **Explain** the formation of an **ox-bow lake**.

You may use a diagram(s) in your answer.   4

Now attempt questions 3 to 6

MARKS

Question 3

Item Q3: Quote from a local landowner

"This area has the potential for a variety of different land uses, including farming, forestry, recreation/tourism, water storage/supply, industry and renewable energy."

Study Item Q3 and the Ordnance Survey map extract (Item A) of the Brecon Beacons area.

Choose **two** different land uses mentioned in Item Q3.

Using map evidence, **explain** how the area shown on the map extract is suitable for your chosen land uses.

5

MARKS

Question 4

Diagram Q4: Synoptic chart, 0800 hours, 10th March

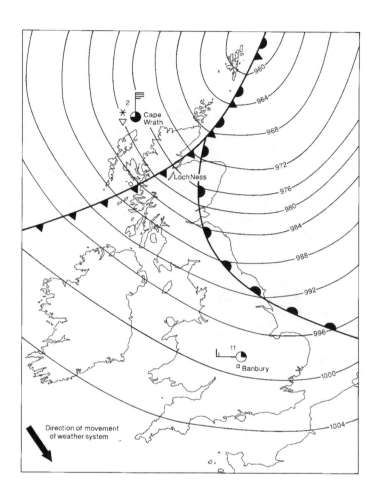

Study Diagram Q4.

(a) **Describe**, in detail, the differences in the weather between Cape Wrath and Banbury at 0800 hours on 10th March.

4

(b) At 0800 hours on 10th March a group of secondary school students are about to set off on a walk into the mountains near Loch Ness. After seeing the weather chart in Diagram Q4, they decide to cancel their walk at the last minute.

Why might conditions have been unsuitable for their expedition? Give reasons.

5

[Turn over

MARKS

Question 5

Diagram Q5: Air masses affecting the British Isles

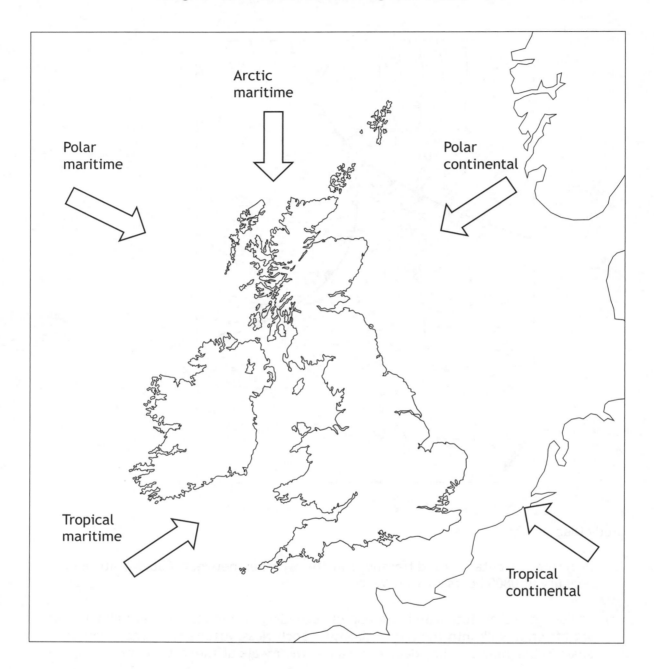

Look at Diagram Q5.

Describe how a long period with a **tropical continental** air mass **in summer** would affect the people of the British Isles.

3

MARKS

Question 6

Diagram Q6: Selected land uses

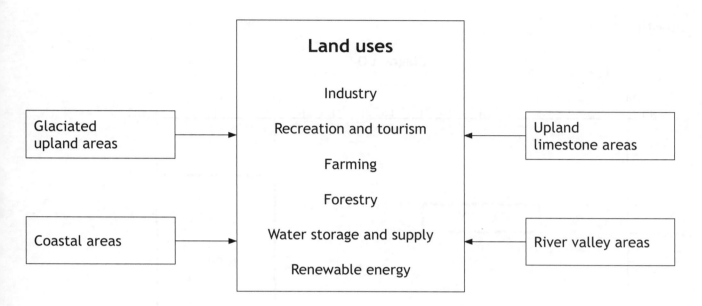

Look at Diagram Q6.

(i) For a named area you have studied, **explain**, **in detail**, ways in which **two** different land uses may be in conflict with each other.

(ii) **Suggest** possible solutions to these conflicts.

5

[Turn over

MARKS

SECTION 2 — HUMAN ENVIRONMENTS — 30 marks
Attempt ALL questions

Question 7

Diagram Q7

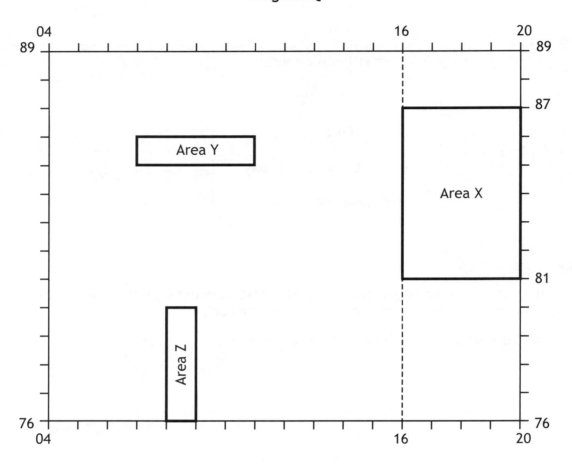

Study the Ordnance Survey map extract (Item B) of the Birmingham area and Diagram Q7 above.

(a) Give map evidence to show that part of the Central Business District (CBD) of Birmingham is found in grid square 0786.

3

(b) Find Area X on Diagram Q7 and the map extract (Item B).

Birmingham airport, a golf course, a business park and a housing area are found in Area X on the rural/urban fringe of Birmingham. Using map evidence **explain** why such developments are found there.

5

(c) The Russell family have three young children and are buying a house in Birmingham. They have narrowed down their search to two areas of the city — Area Y (Balsall Heath and Sparkbrook) or Area Z (Highter's Heath and Drake's Cross).

Which area, Y or Z, should they choose? Using **detailed map evidence, give reasons** to support your chosen area.

6

MARKS

Question 8

Diagram Q8: Developments in farming

GM crops　　　　　　　　　　　　　　　　　　　　　Biofuel

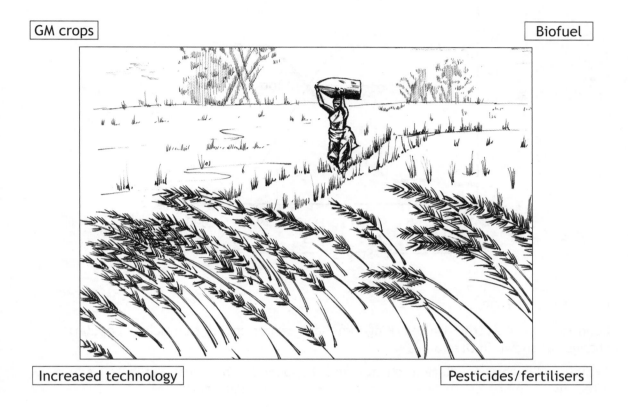

Increased technology　　　　　　　　　　Pesticides/fertilisers

Look at Diagram Q8.

Explain how recent developments in agriculture in developing countries are helping farmers.

4

[Turn over

Question 9   MARKS

Diagram Q9: Demographic transition model

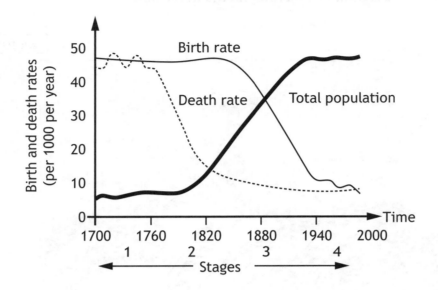

Look at Diagram Q9.

Countries such as the United Kingdom have experienced significant population change, as shown in Diagram Q9.

Explain why this population change has happened. You must refer to factors affecting birth and death rates in stages 2, 3 and 4 of Diagram Q9.   **6**

Question 10

Table Q10: Selected development indicators

Country	Life expectancy (yrs)	Access to safe drinking water (%)	Literacy rate (%)	% of workforce employed in agriculture
Bolivia	69	90	96	32
Chad	50	51	40	80
Finland	81	100	100	4
Mali	56	77	39	80
Netherlands	81	100	100	2
Uganda	55	79	78	40

Study Table Q10.

Choose **two** of the development indicators shown.

For the **two** that you have chosen, **explain**, **in detail**, why they are useful in helping to show a country's level of development.   **6**

MARKS

SECTION 3 — GLOBAL ISSUES — 20 marks
Attempt any TWO questions

MARKS

Question 11: Climate change

Diagram Q11: Area of Arctic Sea ice (1979—2013)

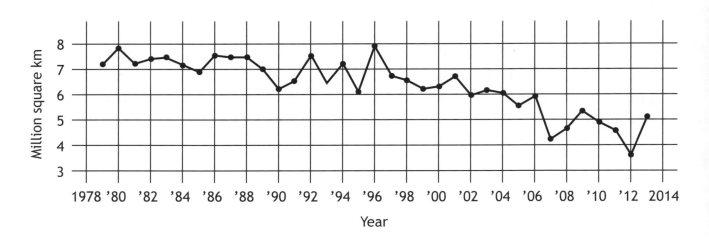

Study Diagram Q11.

(a) **Describe**, **in detail**, the changes in the area of Arctic Sea ice.   4

(b) Melting sea ice is one effect of climate change.

   Explain some other effects of climate change.   6

MARKS

Question 12: Natural regions

Diagram Q12: Recent deforestation rates worldwide

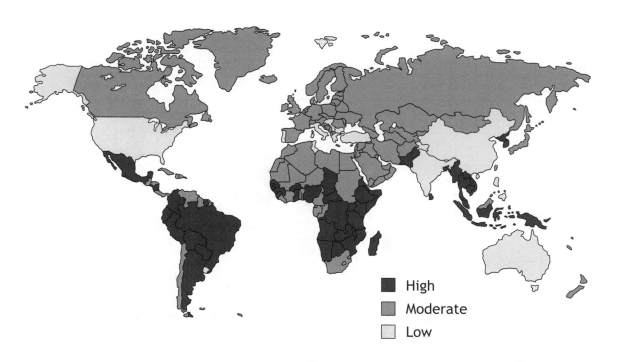

(a) Study Diagram Q12.

Describe, **in detail**, deforestation rates worldwide.   4

(b) **Explain** the management strategies which can be used to minimise the impact of human activity in the tundra.   6

[Turn over

MARKS

Question 13: Environmental hazards

Diagram Q13A: Number of volcanic eruptions per decade 1910—2010

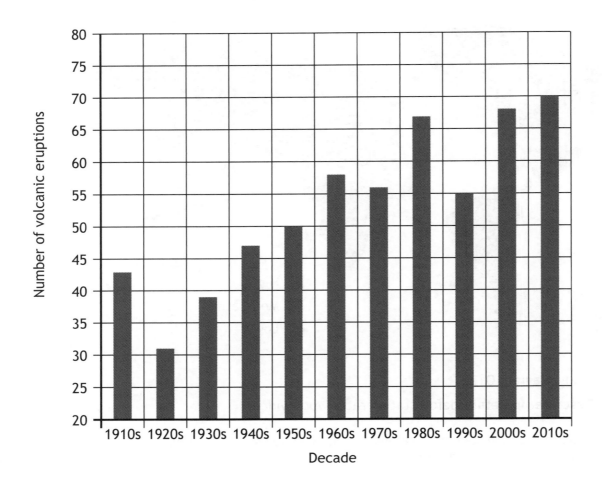

(a) Study Diagram Q13A.

Describe, **in detail**, the changes in the number of volcanic eruptions between 1910—2010.

4

MARKS

Question 13 (continued)

Item Q13B: Pico de Fogo volcano, Cape Verde

> After nearly 20 years of inactivity, the Pico de Fogo awakened with a violent eruption on the 23rd of November 2014.

(b) Look at Item Q13B.

For a volcanic eruption you have studied, **explain**, **in detail**, the impacts of the eruption on people and the landscape.

6

[Turn over

MARKS

Question 14: Trade and globalisation

Diagram Q14A: World exports by region

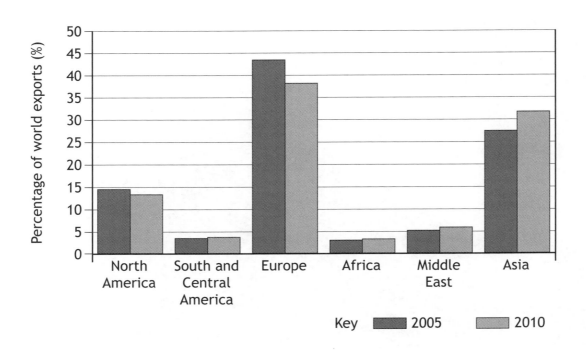

(a) Study Diagram Q14A.

Describe, **in detail**, the changes in world exports from 2005 to 2010.   **4**

Item Q14B: Collecting Fairtrade coffee beans

(b) Look at Item Q14B.

Explain how buying Fairtrade products helps people in the developing world.   **6**

MARKS

Question 15: Tourism

Diagram Q15A: Top ten world tourist destinations (millions of visitors per year)

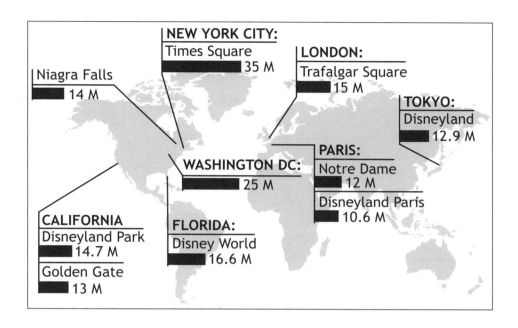

(a) Study Diagram Q15A.

   Describe, **in detail**, the distribution of the top ten world tourist destinations.   4

Item Q15B — Mass tourism on an Italian beach

(b) Look at Item Q15B.

   Describe the effects of mass tourism on people and the environment.   6

MARKS

Question 16: Health

Diagram Q16: Ebola cases in selected African countries April—Oct 2014

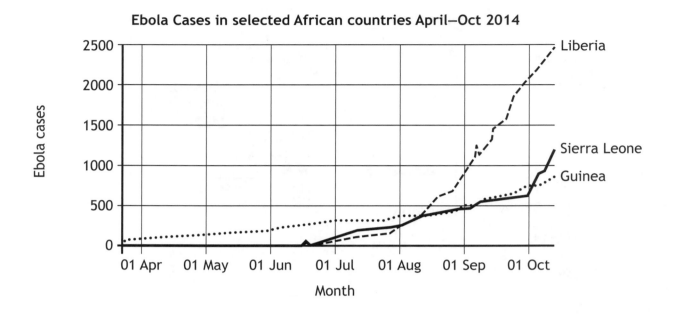

Ebola Cases in selected African countries April—Oct 2014

(a) Study Diagram Q16.

   Describe, **in detail**, the changes in Ebola cases in the **three** named African countries.   **4**

(b) **Explain** the causes of either heart disease **or** cancer **or** asthma.   **6**

[END OF SPECIMEN QUESTION PAPER]

National Qualifications
SPECIMEN ONLY

S833/75/11

Geography
Ordnance Survey Map
Item A

Date — Not applicable

Duration — 2 hours 20 minutes

The colours used in the printing of these map extracts are indicated in the four little boxes at the top of the map extract. Each box should contain a colour; if any does not, the map is incomplete and should be returned to the Invigilator.

Extract No 2213/160

1:50 000 Scale
Landranger Series

Four colours should appear above; if not then please return to the invigilator.

3

ROADS AND PATHS

Not necessarily rights of way

Junction number
Service area Elevated
M1
Unfenced
A 470   Dual carriageway
A 493   Footbridge

B 4518
A 855   Bridge   B 885

	Motorway (dual carriageway)
	Primary Route (recommended through route)
	Main road
	Road under construction
	Secondary road
	Narrow road with passing places
	Road generally more than 4m wide
	Road generally less than 4m wide
	Path / Other road, drive or track
	Gradient: steeper than 20% (1 in 5), 14% to 20% (1 in 7 to 1 in 5)
	Gates, Road tunnel

Ferry P   Ferry V   Ferry (passenger), Ferry (vehicle)

RAILWAYS

	Track multiple or single		Bridges, footbridge
	Track under construction	LC	Level crossing
	Siding		Viaduct, embankment
	Tunnel, cuttings		Station, (a) principal
	Light rapid transit system, narrow gauge or tramway		Light rapid transit system station

WATER FEATURES

Marsh or salting
Towpath   Lock   Slopes   Cliff
Aqueduct   Canal   Ford   Beacon   Flat rock   Shingle   Lighthouse (in use)
Weir   Normal tidal limit   Sand   Lighthouse (disused)
Lake   Footbridge   Dunes   Low water mark
Bridge   Mud
Canal (dry)   High water mark

HEIGHTS

1 metre = 3·2808 feet

50   Contours are at 10 metres vertical interval

·144   Heights are to the nearest metre above mean sea level

Where two heights are shown the first height is to the base of the triangulation pillar and the second (in brackets) to the highest natural point of the hill

ROCK FEATURES

Outcrop
Cliff
Scree

PUBLIC RIGHTS OF WAY

	Footpath
	Bridleway
	Restricted byway
	Byway open to all traffic

The symbols show the defined route so far as the scale of mapping will allow.

The representation on this map of any other road, track or path is no evidence of the existence of a right of way. Not shown on maps of Scotland

Danger Area   Firing and Test Ranges in the area. Danger! Observe warning notices.

OTHER PUBLIC ACCESS

· · · ·   Other route with public access (not normally shown in urban areas). Alignments are based on the best information available. These routes are not shown on maps of Scotland.

●   ●   On-road cycle route
○   ○   Traffic-free cycle route
4   National Cycle Network number
5   Regional Cycle Network number
◆   ◆   National Trail, European Long Distance Path, Long Distance Route, selected Recreational Routes

BOUNDARIES

+ + +	National
+ + +	District
	County, Unitary Authority, Metropolitan District or London Borough
	National Park

ANTIQUITIES

+	Site of antiquity
✗	Battlefield (with date)
☆	Visible earthwork
VILLA	Roman
Castle	Non-Roman

TOURIST INFORMATION

	Camp site / caravan site
	Garden
	Golf course or links
i	Information centre (all year / seasonal)
	Nature reserve
P P&R	Parking, Park and ride (all year / seasonal)
⊼	Picnic site
	Recreation / leisure / sports centre
	Selected places of tourist interest
((	Telephone, public / roadside assistance
V	Viewpoint
V	Visitor centre
!	Walks / Trails
	World Heritage site or area
▲	Youth hostel

LAND FEATURES

⊢ ⊣	Electricity transmission line (pylons shown at standard spacing)
> - > - >	Pipe line (arrow indicates direction of flow)
⌘ ruin	
	Buildings
	Important building (selected)
	Bus or coach station
	Current or former place of worship {with tower / with spire, minaret or dome}
+	Place of worship
	Glass structure
H	Heliport
	Triangulation pillar
⊥	Mast
⊻	Wind pump, wind turbine
⊻	Windmill with or without sails
+	Graticule intersection at 5' intervals
	Cutting, embankment
	Landfill site or slag/spoil heap
	Coniferous wood
	Non-coniferous wood
	Mixed wood
	Orchard
	Park or ornamental ground
	Forestry Commission land
	National Trust (always open / limited access, observe local signs)
	National Trust for Scotland (always open / limited access, observe local signs)

ABBREVIATIONS

Br	Bridge	MS	Milestone
Cemy	Cemetery	Mus	Museum
CG	Cattle grid	P	Post office
CH	Clubhouse	PC	Public convenience (in rural areas)
Fm	Farm	PH	Public house
Ho	House	Sch	School
MP	Milepost	TH	Town Hall, Guildhall or equivalent

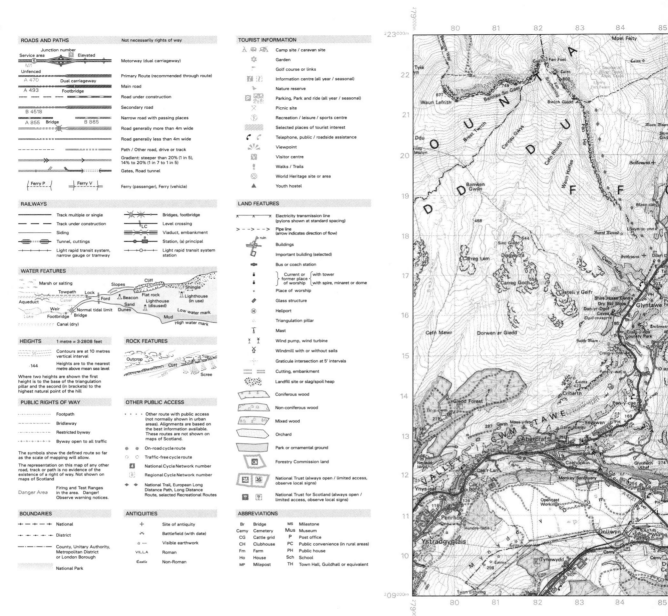

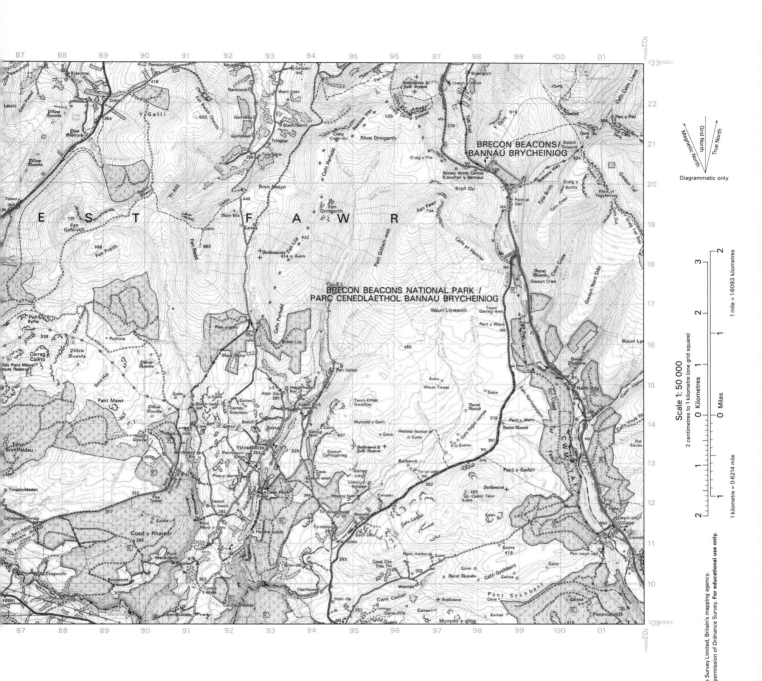

Scale 1: 50 000

2 centimetres to 1 kilometre (one grid square)

[BLANK PAGE]

DO NOT WRITE ON THIS PAGE

National Qualifications
SPECIMEN ONLY

S833/75/11

Geography
Ordnance Survey Map
Item B

Date — Not applicable

Duration — 2 hours 20 minutes

The colours used in the printing of these map extracts are indicated in the four little boxes at the top of the map extract. Each box should contain a colour; if any does not, the map is incomplete and should be returned to the Invigilator.

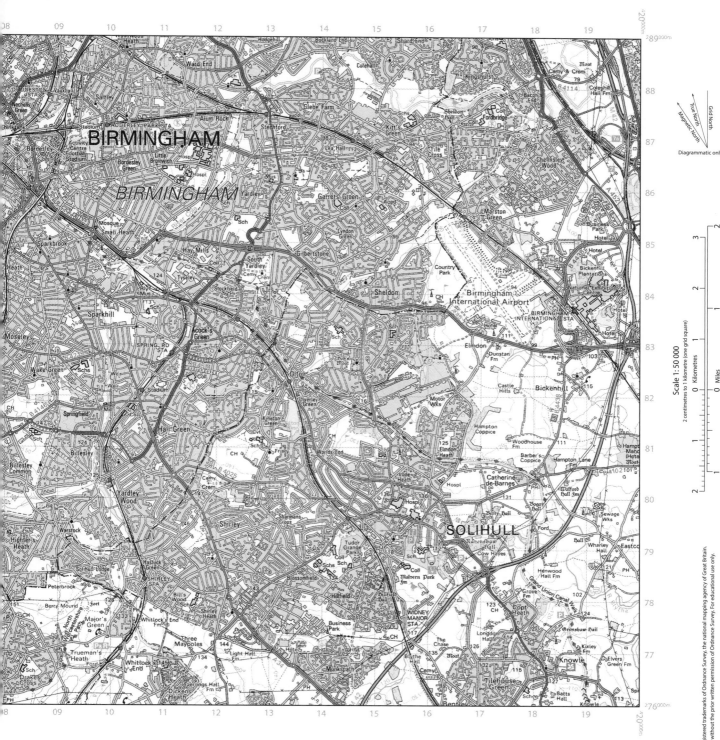

Scale 1:50 000

[BLANK PAGE]

DO NOT WRITE ON THIS PAGE

General Marking Principles for National 5 Geography

Questions that ask candidates to *Describe* . . . (4–6 marks)

Candidates must make a number of relevant, factual points. These should be key points. The points do not need to be in any particular order. Candidates may provide a number of straightforward points or a smaller number of developed points, or a combination of these.

Up to the total mark allocation for this question:

- **One mark** should be given for each accurate relevant point.
- **Further marks** should be given for development and exemplification.

Question: Describe, in detail, the effects of two of the factors shown. (Modern factors affecting farming).

Example:

New technology has led to increased crop yields *(1 mark)*, leading to better profits for some farmers *(second mark for development)*.

Questions that ask candidates to *Explain* . . . (4–6 marks)

Candidates must make a number of points that make the process/situation plain or clear, for example by showing connections between factors or causal relationships between events or processes. These should be key reasons and may include theoretical ideas. There is no need for any prioritising of these reasons. Candidates may provide a number of straightforward reasons or a smaller number of developed reasons, or a combination of these. The use of the command word 'explain' will generally be used when candidates are required to demonstrate knowledge and understanding. However, depending on the context of the question the command words 'give reasons' may be substituted.

If candidates produce fully labelled diagrams they may be awarded up to full marks if the diagrams are sufficiently accurate and detailed.

Up to the total mark allocation for this question:

- **One mark** should be given for each accurate relevant point.
- **Further marks** should be given for developed explanations.

Question: Explain the formation of a U-shaped valley.

Example:
A glacier moves down a main valley which it erodes *(1 mark)* by plucking, where the ice freezes on to fragments of rock and pulls them away *(second mark for development)*.

Questions that ask candidates to *Give reasons* . . . (4–6 marks)

Candidates must make a number of points that make the process/situation plain or clear, for example by showing connections between factors or causal relationships between events or processes. These should be key reasons and may include theoretical ideas. There is no need for any prioritising of these reasons. Candidates may provide a number of straightforward reasons or a smaller number of developed reasons, or a combination of these. The command words 'give reasons' will generally be used when candidates are required to use information from sources. However, depending on the context of the question the command word 'explain' may be substituted.

Up to the total mark allocation for this question:

- **One mark** should be given for each accurate relevant point.
- **Further marks** should be given for developed reasons.

Question: Give reasons for the differences in the weather conditions between Belfast and Stockholm.

Example:
In Stockholm it is dry, but in Belfast it is wet because Stockholm is in a ridge of high pressure whereas Belfast is in a depression *(1 mark)*. Belfast is close to the warm front and therefore experiencing rain *(second mark for development)*.

Questions that ask candidates to *Match* (3–4 marks)

Candidates must match two sets of variables by using their map interpretation skills.

Up to the total mark allocation for this question:

One mark should be given for each correct answer.

Question: Match the letters A to C with the correct features.

Example: A = Forestry *(1 mark)*

Questions that ask candidates to *Give map evidence* (3–4 marks)

Candidates must look for evidence on the map and make clear statements to support their answer.

Up to the total mark allocation for this question:

Question: Give map evidence to show that part of Coventry's CBD is located in grid square 3379.

Example: Many roads meet in this square *(1 mark)*.

Questions that ask candidates to *Give advantages and/or disadvantages* (4–6 marks)

Candidates must select relevant advantages or disadvantages of a proposed development and show their understanding of their significance to the proposal. Answers may give briefly explained points or a smaller number of points which are developed to warrant further marks.

Up to the total mark allocation for this question:

- **One mark** should be given for each accurate relevant point.
- **Further marks** should be given for developed points.
- **Marks** should be awarded for accurate map evidence.

Question: Give either advantages or disadvantages of this location for a shopping centre. You must use map evidence to support your answer.

Example: There are roads and motorways close by allowing the easy delivery of goods *(1 mark)* and access for customers *(1 mark for development)*, eg the A46, M6 and M69 *(1 mark)*.

NATIONAL 5 GEOGRAPHY 2015

Section 1: Physical Environments

1. (a) Headland - 766356
      Cliff   - 690382
      Bay   - 674398

   (b) **Stack**
      Waves attack a line of weakness in the headland (1).
      Types of erosion include hydraulic action, corrosion
      and corrasion (1). Continuous erosion will open up
      the crack and it will develop into a sea cave (1).
      Further erosion of the cave, often on opposite sides
      of the headland, will form an arch (1). The roof of
      the arch is attacked by the waves until it eventually
      collapses (1). This leaves behind a free standing
      piece of rock called a stack which is separate from
      the headland (1).
      Or any other valid point.

      Bay
      Bays are formed due to differential erosion (1) where
      rocks along the coastline are formed in alternating
      bands of different rock types (1) eg sandstone and
      clay (1) and which meet the coast at right angles (1).
      Clay is a softer rock than sandstone so it is eroded
      more quickly (1). The waves erode the softer rock
      through hydraulic action, corrasion and corrosion (1)
      to form sheltered bays (1) which may have beaches
      (1). The harder sandstone areas are more resistant
      to erosion and jut out into the sea to form exposed
      headlands (1).
      Or any other valid point.

2. (a) Levée   - 684466
      Meander   - 708473
      V-Shaped Valley - 713410

   (b) **Meander**
      In the middle/lower course, a river flows downhill
      causing lateral erosion (1). The river contains areas
      of deep water and areas of shallow water, this results
      in areas of slower and faster water movement and
      this causes the current to swing from side to side (2).
      The river flows faster on the outer bank and erodes it
      (1). This forms a river cliff (1). The river flows more
      slowly on the inner bank and deposits some of its
      load (1). This forms a river beach/slip-off slope (1).
      Continuous erosion on the outer bank and deposition
      on the inner bank forms a meander in the river (1).
      Or any other valid point.

      V-shaped valley
      In the upper course, a river flows downhill eroding
      the landscape vertically (1). The river erodes a deep
      notch into the landscape using hydraulic action,
      corrasion and corrosion (1). As the river erodes
      downwards the sides of the valley are exposed to
      freeze-thaw weathering which loosens the rocks and
      steepens the valley sides (2). The rocks which have
      fallen into the river aid the process of corrasion
      which leads to further erosion (1). The river
      transports the rocks downstream and the channel
      becomes wider and deeper creating a V-shaped valley
      between interlocking spurs (2).
      Or any other valid point.

3. Answers will vary depending upon the land uses chosen.

   For farming: Reads Farm (1) (at grid reference 728489) is
   an example of a hill sheep farm as the land is steep (1). As
   the land is higher up, the weather will be harsh and sheep
   can survive these conditions, especially in winter (1). The
   land is too steep for farm machinery to operate (1). The
   soil will be too thin for crops to be grown (1).

   For tourism and recreation: The South West Coastal
   Path follows the top of the cliffs and allows tourists
   to enjoy a view of the coastal scenery (1) eg 727367
   (1). There is a nature reserve for people who want to
   observe wildlife at 747405 (1). There is a golf course
   for golf enthusiasts at 668428 (1). There are various
   camp/caravan sites for people to stay whilst visiting the
   various attractions in the area (1).

4. Answers will vary depending upon the land uses chosen.

   Problems between tourists and farmers:
   In the Cairngorms, tourists can disrupt farming activities
   as walkers leave gates open, allowing animals to
   escape (1). Tourists' dogs can worry sheep if let off their
   lead (1). Stone walls are damaged by people climbing
   over them instead of using gates/stiles (1). Noisy tourists
   can disturb sheep especially during breeding season (1).
   Farmers may restrict walkers access at certain times eg
   lambing season (1). Farm vehicles can slow up tourist
   traffic on roads (1) and parked cars on narrow country
   roads can restrict the movement of large farm vehicles (1).

   Problems between industry and tourists:
   Tourists want to see the beautiful and unusual scenery of
   the Yorkshire Dales but quarries spoil the natural beauty
   of the landscape (1). Lorries used to remove the stone
   endanger wildlife and put visitors off returning to the
   area (1). This threatens local tourist-related jobs eg in
   local restaurants (1). The large lorries needed to remove
   the quarried stone cause air pollution which spoils the
   atmosphere for tourists (1). Lorries cause traffic congestion
   on narrow country roads which slows traffic and delays
   drivers (1). The peace and quiet for visitors is disturbed by
   the blasting of rock (1). Some wildlife habitats may also be
   disturbed by the removal of rock (1).
   Or any other valid point.

5. South-East England is usually warmer because it is closer
   to the Equator (1). This is due to intense heating from
   the sun (1) because sun rays are more concentrated
   (1). Places in Northern Scotland eg Wick, are colder
   because they are closer to the North Pole (1). This is due
   to a lack of insolation from the sun as the rays are less
   concentrated (1) and reflection of heat by the snow and
   ice (1). Places located on flat low-lying land are warmer
   eg Central Scotland, because temperatures increase as
   altitude decreases and places higher up ie mountainous
   regions are colder (1) because temperature decreases
   by 1°C for every one hundred metres in height (1).
   Places which are south facing are warmer because they
   get more sun (1) and places which are north facing are
   colder because they experience cold northerly winds (1).
   Western coastal areas are warmer because of a warm
   ocean current (1) (The North Atlantic Drift) and due to
   the prevailing South-Westerly winds that are warmed as
   they pass across this warm ocean current (1). In summer,
   places closer to the sea are cooler and in winter they are
   warmer because the sea heats up slowly in summer and
   cools slowly in winter (2).
   Or any other valid point.

Section 2: Human Environments

6. (a) Main roads lead into this square (1) there is a bus station (1) and two railway stations (1) tourist information centre (1) several churches (1) museum (1). Or any other valid point

(b) The land is flat so easy to build on (1) there is space available for expansion (1) eg expansion of the motor works at 163823 (1). There are good transport links like the M42 allowing people and products access to and from the area (1). A rail link with Birmingham International Rail Station gives easy access to the airport (1). There are many road junctions and intersections connecting the area to other areas and less traffic congestion as it is away from Birmingham city centre (2). The land is on the edge of Birmingham so will be cheaper encouraging housing estates like Sheldon to be built (1). The cheaper land allows the houses to be bigger with cul-de-sacs, gardens etc.(1). The houses can provide a source of labour for the airport, motor works and the business park (1).
Or any other valid point

7. Contraception and family planning is widely available (1). Later marriages are more common which results in fewer children (1). People no longer choose to have lots of children as improved medical care and advances in medicine (1) have resulted in most children surviving at birth (1). Developed countries have the money to invest in medical care which reduces the infant mortality rate thus causing the birth rate to fall (1). Children are expensive so the greater number of children the bigger the financial burden (1). Women want careers so put off having children to a later age (1) or limit the size of their families to give them a reasonable standard of living (1). Sex education in schools helps to lower birth rates (1).
Or any other valid point.

8. For example in Rocinha (Rio), the former wooden shacks have been upgraded to permanent dwellings with some modern services (1). Residents constantly improve their homes through a process of 'self-help' (1) where the residents are provided with materials like bricks (1). Some prefabricated houses have been built by the Brazilian government (1) with basic facilities like toilets, electricity and running water (1). The residents have been given the legal rights to the land (1), roads have been built into/or improved in the favela (1) allowing services like rubbish collections to take place (1), there are now a few health clinics and schools provided (1).
Or any other valid point.

Section 3: Global Issues

9. (a) The overall trend is that the amount of Arctic Sea ice has decreased between 1979 and 2013 (1) from (around) 7 million square kilometres to (about) 5 million square kilometres (1). There has been a fluctuation in the extent of sea ice in certain years (1) eg in 2013, the amount of sea ice increased from 3.75 million square kilometres in 2012 to 5 million square kilometres (1) whereas between 2006 and 2007 there was a sharp decrease (1) from 6 million square kilometres to 4.25 million square kilometres (1).
Or any other valid point.

(b) Increased temperatures are causing ice caps to melt so Polar habitats are beginning to disappear (1). Melting ice causes sea levels to rise (1) threatening coastal settlements (1). An increase in sea temperatures causes the water to expand, compounding the problem of flooding (1). Global warming could also affect weather patterns, leading to more droughts (1) crop failures and problems with food supply (1); flooding, causing the extinction of species (1) and more extreme weather, eg tropical storms (1). Tourism problems will increase as there will be less snow in some mountain resorts (1). Global warming could threaten the development of developing countries as restrictions on fossil fuel use may be imposed to slow the rate of increasing CO_2 levels (1). In the UK, tropical diseases like malaria may spread as temperatures rise (1). Plants growth will be affected and some species will thrive in previously unsuitable areas (1). Higher temperatures may cause water shortages (1).
Or any other valid point.

10. (a) Overall the amount of deforestation in Peru 2004–2012 has decreased (1) from just under 3 million ha to 750 000 ha (1). The deforestation rate declined rapidly from 2004 to 2007 (1). Deforestation increased from 2007 to 2008 peaking in Peru at 1 500 000 hectares per year (1). Again Peru experienced a decline in deforestation rates from 2008 to 2009 by over 500 000 ha (1). From 2009 to 2010 deforestation rates rose to around 1 400 000 ha (1), before declining to around 750 000 hectares per year in 2012 (1).

(b) New industries have led to the expansion of towns such as Anchorage in Alaska which have grown to accommodate workers (1). Although these industries provide employment (1), these developments spoil the appearance of the natural landscape (1). New roads have been built to transport people and goods. This increases the number of vehicles in the tundra creating noise and air pollution (1). But also improves access to locals (1).

Oil is a very important industry in Alaska. The building of oil platforms and oil pipelines has resulted in damage to tundra vegetation and wildlife (1). In some areas, the Trans-Alaskan oil pipeline has been built on natural migration or hunting routes for animals, which hinders the natural movement of caribou (1). Local Inuit people have also had their way of life disrupted as they must detour around the pipeline (1) and may no longer have access to their traditional hunting grounds (1).

Local people were promised jobs in the industry, but few jobs are available for locals (1).

Burst pipes have spilt hundreds of thousands of gallons of crude oil in Alaska, devastating this fragile environment (1). Oil spills have also been responsible for pollution in the region (1), such as the Exxon Valdez disaster (1).

Any damage to the tundra landscape is slow to recover, as the short growing season means that bulldozer tracks from the oil and natural gas industries could take centuries to restore (1).

Pollution from mining and oil drilling has contaminated the air, lakes and rivers (1).
Or any other valid point.

11. (a) Most cities are located on or near plate boundaries (1) where seismic activity is highest (1). Most earthquake threatened cities are found in developing countries (1) like Indonesia (1). A large number of threatened cities are found in China (1). Three cities in Africa are at risk (1). All threatened cities in the USA are found on the west coast (1) with a cluster around San Francisco/Los Angeles (1).
Or any other valid point.

(b) In Japan people take part in earthquake drills to practise what to do in the event of an earthquake (1) giving them a better chance of survival (1). The government warn people, using text messages and TV, giving them the chance to move to a safer place (1). Earthquake resistant buildings reduce the number of people trapped or killed (1). Sprinkler systems and gas cut off valves prevent fires spreading reducing the number of people injured and buildings destroyed (1). People living in earthquake prone areas have emergency plans in place and emergency supplies such as bottled water and tinned food are stockpiled to ensure they have vital supplies to survive in the event of an earthquake (2). In the event of an earthquake short term aid in the form of food, medicine and shelter is sent to the area to treat the injured (1).
Or any other valid point.

12. (a) The value of exports from developed world countries to developing world countries is $738bn (1) whereas there is only $650bn worth of goods exported from developing to developed world countries (1). That is a difference of $88 billion (1). The value of trade between developing world countries is $383bn (1). The value of trade between developed world countries is $2251bn (1). There is more trade between developed world countries than between developing countries (1); it is $1868bn more (1).
Or any other valid point.

(b) There is a big imbalance in the pattern of trade between the developing and developed world; this can reinforce differences in wealth between areas such as the EU and Africa (1); African countries export mainly primary products such as oil or cocoa beans for comparatively low prices but import mainly processed goods such as vehicles for much higher prices (1) which can result in a trade deficit for them (1); this can increase levels of poverty within African countries and cause difficulties for the economy as well (2); often the producers such as cocoa farmers in Africa receive very low wages and so struggle to maintain a decent standard of living (2); wealthy European countries profit from selling expensive manufactured goods to African countries (1), helping to keep a much higher standard of living for their citizens (1); often, exploitation of primary products in African countries can lead to serious environmental damage, such as logging which has caused deforestation (1), resulting in the loss of areas of rainforest as well as the destruction of animal habitats (1).
Or any other valid point.

13. (a) There has been a fairly steady increase in visitor numbers since 1995 (1) from around 525 million reaching 1 billion in 2013 (1). There were only 2 years where the numbers decreased slightly ie in 2003 (1) when it dropped to just under 700 million (1) and in 2009, dropped to under 900 million (1). The period with the largest increase was the 5 years between 1995 and 2000 (1) whereas the slowest increase has been in recent years from 2010 (1).

(b) If Costa Rica cloud forest chosen:
Eco-tourism raises local as well as international awareness of natural environment (1) such as wildlife and vegetation (1). Developing countries now want to conserve their fragile environments and view eco-tourism as a significant means of generating income (1). Developed countries want to help developing countries conserve their fragile environments by promoting sustainable/eco-tourism (1). Tourists are now more environmentally conscious and want to help protect fragile environments for future generations (1). Eco-tourism provides work and opportunities for local people (1) hence improving their standard of living (1) encourages local enterprise and improvement schemes (1) promoting awareness of local culture and traditions (1).
Or any other valid point.

14. (a) Male deaths from heart disease are most common in Eastern Europe (1). Russia for example, has a rate of 444–841 per 100 000 (1). This compares to only 120–238 in the UK (1). Canada, the USA and Mexico have some of the lowest rates (1), with under 120–238 per 100 000 (1). Many central African countries have rates of 363–443 (1).
Or any other valid point.

(b) If **pneumonia** chosen:
Antibiotics are used to treat any bacterial lung infections (1) and patients are encouraged to drink plenty in order to avoid dehydration (1); in severe cases a drip may be required to restore the right level of salts and fluids quickly (1); paracetamol is used to ease the effects of fever and/or headaches (1); introducing more community-based health workers helps to control the incidence of pneumonia as children with the disease are more likely to be diagnosed and treated quickly (1); this can often help to save lives (1). Vaccinations are being increasingly used in developing world countries to protect children against common infections such as flu (1); adequate nutrition helps to increase a child's natural defences against disease and so education about this also helps to reduce pneumonia (1).
Or any other valid point.

If **kwashiorkor** chosen:
The main method of managing kwashiorkor is education about the need for a well-balanced diet, so that children don't develop the disease in the first place (1); by educating communities they can be encouraged to grow different food types to increase protein intake (1); this might include crops

such as cashews, peanuts, lentils or sunflower (1) and might also involve advice about constructing irrigation schemes to help crops grow better in times of drought (1); education about family planning also helps to reduce the number of children per family, making more food available per child (1).

For children who have kwashiorkor it is important to give vitamin and mineral supplements as salt and mineral levels in their blood stream may be dangerously low (1); Zinc supplements might also be administered to help the skin recover (1). Small amounts of food are reintroduced slowly, such as carbohydrates to give energy (1) and protein rich foods to help the child's body recover (1).
Or any other valid point.

If **malaria** chosen:
Anti-malarial drugs kill blood parasites (1). Chloroquine is an example of this (1). Insecticides, such as malathion destroy the female anopheles mosquito (1).

Draining all breeding areas eradicates larvae (1), planting eucalyptus trees to soak up moisture removes breeding ground (1). Water can also be released from dams to drown immature larvae (1). Mustard seeds can be used to drag larvae below the surface to drown them (1). Small fish can be introduced to eat larvae and provide a cheap protein source (1). Genetic engineering of sterile male mosquitoes reduces mosquitoes (1).

Health education teaches people about how to protect themselves from being bitten (1). Preventative bed nets are cheap and effective at stopping mosquitos biting at night (1). New treatments have also been developed which seem to be more effective such as artemesinin/ACT because malaria parasite is not yet resistant to them (1).
Or any other valid point.

If **cholera** chosen:
One of the main ways to reduce or control the spread of cholera is to improve sanitation which stops disease from spreading (1). Providing wells and pipes makes drinking water safe and clean (1). Health Education encourages people to wash hands often with soap and safe water preventing infection as does building and use of latrines (2). Because of contaminated water people should cook their food well and eat it hot (1). Food stuffs should be kept covered and fruit and vegetables should be peeled to prevent contamination (2).

Cholera is an easily treatable disease.
The main ways to treat cholera are either a simple drink made from 1 litre of safe water, 6-8 teaspoons of sugar and 1/2 teaspoon of salt, which helps to rehydrate sufferers so that they can fight off the disease (2) or re-hydration tablets, if available (1). In especially severe cases, intravenous administration of fluids may be required to save the patient's life (1). Treatment with antibiotics is recommended for severely ill patients to help fight the infection (1).
Or any other valid point.

NATIONAL 5 GEOGRAPHY 2017

Section 1: Physical Environments

1. (a) A = arête
   B = corrie
   C = U-shaped valley

(b) Snow compresses to ice and forms a glacier (1). The glacier uses the process of plucking to steepen the sides of the valley (1). Plucking is when ice sticks onto rocks at the sides of the valley and as the glacier moves downhill, it rips the rocks out (1). Abrasion happens when rocks frozen into the base of the glacier grind at the valley floor as the glacier moves (1). The glacier uses the process of plucking and abrasion to widen and deepen the valley (1). The valley is also weathered above and below the glacier by frost shattering (1). Interlocking spurs are cut-off by ice creating truncated spurs (1).

Or any other valid point.

(c) **Farming:** Hill sheep farming is common in a glaciated upland area such as the Cairngorms because sheep are hardy and can survive the cold, harsh conditions (1). The low temperatures and lack of sunshine mean the climate is unsuitable for growing crops (1). Crops are also unable to grow as high rainfall leaches nutrients from the soil (1). The slopes are too steep to use farm machinery (1). Flatter areas on valley floors are often marshy which makes it unsuitable for arable farming (1). Some pastoral farming is possible on valley floors as the grass is better quality (1).

Forestry: Commercial forestry can take place on the lower slopes of u-shaped valleys where weather conditions are less harsh and soil quality is better (1). This is possible as trees are hardy and can grow on quite steep land and relatively thin soils (1). Trees make use of steep land that is unsuitable for farming or building on (1). Trees help to prevent soil erosion on slopes and flooding in valleys as their roots bind soil together and absorb water (2).

Recreation and Tourism: Tourists are attracted to glaciated upland areas for the natural scenery which includes ancient forests, vast mountains with glacial features, rivers and lochs (1). Ribbon lochs provide opportunities for water sports such as water skiing and canoeing (1). Mountains provide great opportunities for hill walking and rock climbing (1). Snow-filled corries enable winter sports such as skiing and snow-boarding (1). Bird watching is also popular in forests (1). Small settlements eg Aviemore provide tourist services such as hotels, eateries, information centres/car parks/equipment hire shops (1).

Water Storage and Supply: The high rainfall in upland areas supplies lochs with water that can be used to provide drinking water to settlements (1). The hard impermeable rocks provide excellent geological conditions for water storage in reservoirs (1). Steep sided u-shaped valleys provide a natural basin for water storage (1).

Renewable Energy: Hydro-electric power (HEP) is generated by damming hanging valleys to create electricity using the force of waterfalls (1). Wind turbines can also be located on mountains to take advantage of the windy conditions to generate energy (1).

Or any other valid point.

2. (a) A = stalactite
       B = grike
       C = joint

   (b) Limestone is made from the decayed remains of sea creatures laid in horizontal layers on sea beds (bedding planes) (1). These sedimentary rocks were uplifted (1) and cracks appeared as the rocks dried out (joints) (1). During glaciation, ice scraped away the topsoil and exposed the bare rock underneath (1). The dry, well-jointed (permeable) bare rock surface allows water to seep down into it (1). Acidic rainwater reacts with the limestone and dissolves the rock (carbonation) (1). The dissolved limestone is carried away by running water (solution) (1). Continued chemical weathering widens and deepens cracks to form gaps called **grikes** (1). Rectangular blocks of limestone called **clints** are separated by the grikes (1).

       Or any other valid point.

   (c) **Farming:** Hill sheep farming is common in upland limestone areas such as the Yorkshire Dales because sheep are hardy and can survive the harsh weather conditions and poor quality grazing (1). Some dairy farms are located on the flat land in the valleys where the soil is more fertile to provide better quality grazing and the weather is warmer and drier (2). A lack of surface water, thin soils and bare rock mean that crops cannot be grown (1).

       Industry: Quarrying is often an important industry in upland limestone areas (1). In the Yorkshire Dales, the main rocks quarried are carboniferous limestone, sandstone and gritstone (1). Cement works can also locate in limestone areas for the raw material lime (1).

       Recreation and Tourism: Tourists visit limestone areas to see the distinctive landscape eg limestone pavements, scars and potholes (1). Visitors like to enjoy the experience of traditional idyllic rural villages (1). People visit limestone caves eg White Scar Caves in the Yorkshire Dales to admire the dripstone features (1). Hill walking in the uplands and cycling in the valleys are common activities (1). Abseiling down limestone scars is a popular activity (1). Many other activities such as caving, pot-holing, rock climbing and horse riding are also popular in limestone areas (1).

       Renewable Energy: Upland areas are suitable for generating wind power as they are higher up so more exposed to wind (1).

       Or any other valid point.

3. It is –2C because it is a high pressure/winter anticyclone (1) and there is often a lack of cloud allowing heat to escape, bringing low temperatures (1). The temperature is very cold because it is December in the UK ie winter time (1). Winds will be gentle as isobars are widely spaced (1). Wind direction is westerly as winds blow clockwise around anticyclones in UK (1). There is little cloud (1 okta) as cold air sinks in an anticyclone (1). The weather is dry as there are no fronts to bring rain/snow (1).

   Or any other valid point.

4. **Advantages:** Warm, dry and sunny weather improves people's mood (1). People can participate in more outdoor activities such as BBQs (1). Outdoor sports can take place eg tennis matches without being rained off (1). School sports days can safely go ahead due to dry conditions (1). Rising sales of summer goods such as sunscreen and ice lollies increase shops' profits (1).

Disadvantages: Hose-pipe bans enforced due to lack of water (1). Drought conditions reduce the yield of farmers' crops (1). People suffer from sunburn and dehydration (1). More people admitted to hospital with heatstroke (1) putting a strain on resources (1). Forest fires break out (1). Thunderstorms are also a disadvantage of anticyclones (1).

Or any other valid point.

Section 2: Human Environments

5. CBD   - 2573
   Old housing   - 2671
   New housing - 2568

6. There is flat land to easily build the houses on (1). The land on the rural/urban fringe is cheaper, so low density housing with gardens/garages can be built (1). There is good road access to this area via the A720/A701/B701 (1) which people can use to commute to their work (1). The area is on the edge of the city, so there will be less noise and air pollution (1) and less traffic, so it will be safer for families (1). There are woods nearby, where residents can go for walks to relax (1). There are also other good opportunities for outdoor recreation such as the ski centre and country park at Hillend (1). There is a Park and Ride scheme nearby which gives easy access to the city (1).

Or any other valid point.

7. **PHYSICAL FACTORS**

   Relief: People prefer to live on flat, low-lying areas because it is easier to build on (1). Coastal areas allow trade to take place as ports locate by the sea so many people live nearby (1). Few people tend to live in mountainous areas because steep slopes make it difficult for machinery to operate (1). Upland areas are too cold and wet which makes it difficult to grow crops, so few people live there (1). Mountainous areas also have a low population density because they are often isolated which makes them hard to access (1).

   Climate: Many people prefer to live in temperate climates where there is enough rainfall to provide water (1). Few people tend to live in areas with extreme climates because areas like the Sahara Desert with very high temperatures and low rainfall make farming difficult (1). Few people live in areas such as Arctic Canada as permafrost makes building houses and roads difficult as the ground is frozen for much of the year (1). Rainforests have a low population density as they are uncomfortable to live in due to the humid climate (1) and diseases like Malaria spread easily (1).

   Soil: People prefer to live in areas with fertile soils so that crops can be grown to supply food (1). Where there are poor quality soils eg on steep slopes in Northern Scotland, few crops can be grown so less people live there (1). Few people live in hot desert areas because soil dries out and turns to dust, making it difficult to grow crops/keep animals (1).

   Natural Resources: Many people tend to live in areas where there are minerals and raw materials to extract and sell (1). Natural landscapes with beautiful scenery attract tourists which generates job opportunities (1) in hotels, shops and restaurants, so more people live in those areas (1). Few people tend to live in areas lacking natural resources because there will be little industry and this means less employment opportunities (1).

HUMAN FACTORS

Job Opportunities: Jobs in different industries in urban areas encourage people to move to find work (1). Cities such as Rio de Janeiro have a high population density as there are a variety of job opportunities (1).

Transport and Communications: Areas which are more accessible eg Central Lowlands of Scotland tend to have higher population densities (1). Places with good transport links attract people and industries which in turn creates employment opportunities, so more people live there (1).

Services: Towns and cities are crowded as people move to cities like Berlin, London and New York for a variety of amenities and services eg education, health care, jobs and entertainments (1).

Government Aid: Industries locate where there is government funding available, as a result, people move into these areas for work (1). Population density in areas like Syria is falling as people are moving away because of prolonged war (1).

Or any other valid point.

8. For example: In Rio de Janeiro: Wooden shacks have been upgraded to permanent dwellings with some services (1). For example, clean piped water has been provided to help reduce the spread of diseases (1). Residents continually upgrade their homes through a process of 'self-help' schemes (1) where the local people are provided with materials like bricks (1). Some prefabricated houses have been built in Rocinha by the Brazilian Government (1) with basic facilities like toilets and electricity (1). The residents have been given legal rights to the land where their house is built (1). Roads have been built/improved in the favela allowing services like rubbish collections to take place (1).

In some favelas cable car systems have been constructed to improve transport for residents (1). There have been some schools and health clinics provided for residents (1). Some charities have also donated money to help improve the standard of living of people in shanty towns (1) eg by providing computers in schools (1). Security has been improved by having more police patrols (1) which have helped to reduce drugs related crime (1).

Or any other valid point.

Section 3: Global Issues

9. (a) Overall the temperatures increased between 1996–2016 (1) by 0.6 degrees (1). The biggest increase was between 1997–1998 where it increased by 0.15 degrees (1). Temperatures decreased only five times in 20 years (1) with the largest decrease between 1998–1999 where it dropped by 0.2 degrees (1). It took seven years for temperatures to reach the same level increase as 1998 at just over 0.5 degrees (1). Temperatures have been continually rising since 2011 reaching the highest level in 2016 with an increase of just over 0.4 degrees (1).

Or any other valid point.

(b) In the UK the government encourages people to make their houses more energy efficient by giving grants for things like loft insulation, which reduce the amount of energy used (1). Turning off lights, electrical appliances and turning down thermostats reduces the amount of fossil fuels used putting less CO_2 into the atmosphere (1). Many countries like the UK encourage the use of public transport so reducing the damaging emissions from cars (1). In Brazil laws have been passed to reduce the removal of forest through burning and illegal logging so reducing the amount of CO_2 released into the atmosphere (1). The UK government is trying to reduce the use of fossil fuels such as coal, oil and natural gases by introducing targets for renewable energy using green fuels such as HEP, wind power, solar power (1). Many world nations including the UK take part in Climate Change Conferences, for example the Paris Conference December, 2015 where nations agreed targets to reduce the causes of global warming (1). An increasing number of cities are introducing policies to reduce car use and therefore greenhouse gas emissions (1) such as the new tram system in Edinburgh (1) and bicycle friendly infrastructure in Amsterdam (1).

Or any other valid point.

10. (a) The graph for Canada shows a tundra climate while the graph for Indonesia shows a tropical rainforest (equatorial) climate (1). The tropical rainforest climate is much wetter and much warmer than the tundra climate (1). The wettest month in the rainforest has about 310 millimetres of rain whereas the wettest month in the tundra has about 30 millimetres (1). January and March are the wettest months in the rainforest whereas July is the wettest month in the tundra with some months having less than 10 millimetres (2). The highest temperature in the rainforest is about 27 degrees C but in the tundra it is only 4 degrees C (1). The range of temperature in the tundra is 38 degrees C but only 2 degrees C in the rainforest (1).

Or any other valid point.

(b) **Environment:** In Indonesia for example, multiple fires have destroyed large areas of the tropical rainforest (1), completely destroying animal habitats and the entire ecosystem (1). Orang-utans have been particularly badly affected by this as they are already endangered and have been further threatened by the destruction of their habitat (1). Other animals such as tigers have fewer animals to prey on due to the smaller natural areas of forest which are left and often impact on local communities by taking their livestock instead (2). Many of the fires have been caused by small scale farmers who want to clear areas of trees to plant them with cash crops such as palm oil (1). Pollution has resulted from burning rainforests and has been so bad that cities such as Jakarta and Kuala Lumpur have been covered in a thick smog, partly caused by rainforest burning (1). The fires also add large amounts of carbon dioxide to the atmosphere leading to further global warming and climate change (1). Road building through the forest not only destroys all the vegetation but opens up new areas to exploitation by small scale farmers or by logging companies so the construction of roads has an especially bad effect on the forest (2).

People: All of these activities impact also on indigenous peoples who have lived in the forest for generations — they lose their ancestral lands, their food sources, culture and way of life (2). Also when they come into contact with outsiders, they may contract illnesses to which they have little or no resistance, resulting in serious illness or worse (1). Sometimes indigenous communities have had their lands forcibly taken from them with outbreaks of violence resulting in many casualties among the forest peoples (2).

Or any other valid point.

11. (a) Tropical storms (also called hurricanes, typhoons and tropical cyclones) form over oceans within 30° North and South of the equator (1) generally where sea temperatures rise over 27°C (1). They are known as hurricanes where they form over the Atlantic Ocean heading westwards towards the Caribbean/the east coast of Central America/Southern USA eg Florida (1). Tropical cyclones form in the Indian Ocean and move towards Bangladesh/Pakistan/India/Indian Ocean islands such as Mauritius/Madagascar (1). Typhoons form in the Pacific Ocean and South China Sea and affect Australia and countries in South East Asia such as the Philippines, China and Japan (1).

Or any other valid point.

(b) **People:** (eg in Typhoon Haiyan which struck the Philippines in November 2013). The subsequent high seas and flooding resulted in over 6,000 people being killed (1). Whole communities and buildings were destroyed by the intense winds of over 196 mph (1). Over a million people were made homeless and suffered from stress due to loss of possessions and housing (1). Roads and railways were destroyed leading to communication problems and making rescue efforts almost impossible (2). Electricity lines were blown down and people were without power supplies for months (1). People were stranded due to flooding which would have been traumatic (1). Fishing boats and other craft may be damaged causing loss of income (1). There were food and water shortages which led to ill-health (1). As a result of extensive flooding, people may catch water-borne diseases which could be fatal (1). There may be looting of homes, factories and other properties causing tension (1). People lost their jobs in factories that had been destroyed (1). Insurance claims are made resulting in the cost of insurance premiums rising in the future (1). Whilst businesses are closed, earnings (and profits) will be lost (1). Crops were damaged which led to lower productivity and loss of earnings from exports (1). Even after three years thousands of people in cities such as Tacloban are still living in temporary accommodation, reducing their quality of life (1).

Environment: Extensive flooding occurs as a result of the huge amount of rain which falls during a tropical storm (1). Flooding can lead to sewer systems overflowing and spreading disease (1). There will be structural damage to buildings which may have to be pulled down and rebuilt (1). Sensitive ecosystems may be destroyed and plant and animal habitats lost (1). Fish are often killed in storm surges and because of silting (1). Crops and livestock may be damaged or completely destroyed (1). After Typhoon Haiyan, mudslides were common because the soil was saturated (1). They flowed quickly down hillsides destroying houses and crops and killing people and livestock (1). In many parts of the Philippines, extensive coastal erosion resulted in loss of farmland and whole communities (1).

Or any other valid point.

12. (a) Between 1995 and 2010 the developing countries' share of world trade has increased, while the developed countries' share has gone down (1). In 1995 developed countries accounted for just under 70% of world trade, but by 2000 this had dropped to around 68% (1). By 2010 developed countries share had dropped again by around 13% (1).

In 1995 developing countries accounted for just 28% of world trade, but by 2000 it had risen to around 31% and by 2010 it rose again by 9% (1).

Or any other valid point.

(b) Trade is the exchange of goods and services between countries. More than half the world's trade takes place between just eight countries known as the G8 (1). Usually, developed countries export valuable manufactured goods such as electronics and cars and import cheaper primary products such as tea and coffee (1). In developing countries, the opposite is true. This means that developing countries have little purchasing power, making it difficult for them to pay off their debts or escape from poverty (2). The price of primary products fluctuates on the world market. Workers and producers in developing countries lose out when the price drops, but they benefit when it rises (2). This instability makes it difficult to plan improvement, either locally on farms or in wider government (1). Sometimes developed countries impose tariffs and quotas on imports (1). Tariffs are taxes imposed on imports, which makes foreign goods more expensive to the consumer (1). Quotas are limits on the amount of goods imported and usually work in the developed country's favour (1). Poorer countries supply resources such as timber, agriculture, oil and mining products, often at low prices. These products are used in manufacturing industries to make products which are then sold for large profits, often to poorer countries (2). Often poor countries rely on only one or two raw materials such as Ecuador which grows bananas (1). When the price or demand for bananas falls, the country's income can be badly affected (1). This means countries may need to turn to borrowing and increasing their debts (1).

Or any other valid point.

13. (a) Between 2006 and 2014 the number of tourists visiting Scotland from the USA decreased from 475,000 in 2006 to 275,000 in 2010 (1) and then increased to 418,000 in 2014 (1). The number of tourists coming to Scotland from France, USA, Canada, Ireland and Spain decreased (1). There was an increase in tourists visiting Scotland from Germany, Australia, the Netherlands and Scandinavia (1). The amount of tourists from the Rest of the World increased from 976,000 in 2006 to 1,106,000 in 2014 (1). The number of people visiting Scotland from Australia increased over the years from 133,000 in 2006 to 158,000 in 2014 (1). Overall, the total number of tourists visiting Scotland decreased from 2,732,000 in 2006 to 2,700,000 in 2014 (1).

Or any other valid point.

(b) Mass tourism has increased due to improvements in road, rail and air travel which enables people to travel more easily (1). Holiday pay means people can afford to take time off work for a break (1). Increased time off work eg Bank holidays gives people the opportunity to visit different places (1). Tour operators and travel agents make it easier to go on holiday abroad due to package deals which often include flights, transfers, meals and holiday reps on hand to solve any problems (2). Cheap package holidays and budget airlines such as EasyJet make holidays more affordable to many people (1). TV travel programmes and adverts on social media inspire people to visit foreign locations (1). The demand to explore various places of interest has increased as globalisation has made the world smaller (1). People also want to experience different cultures and new adventures (1).

Or any other valid point.

14. (a) Mostly there has been a global reduction in worldwide mortality rates from malaria (1), although in South America countries such as Surinam and Venezuela have experienced an increase since 2000 (1). In Brazil and Peru, the mortality rates have decreased by 75% since 2000 (1) and in Argentina the disease appears to have been eliminated since 2000 (1). In Africa most countries have experienced a reduction of between 0% and 74% (1). Nigeria and Kenya are both in this category (1). In SW Africa, countries such as Namibia and Botswana have had a reduction of over 75% in malaria mortality (1).

Or any other valid point.

(b) If **malaria** chosen:

Malaria happens when the parasites injected into the bloodstream by mosquitoes migrate to the liver, multiply and break out in a new form to attack the red blood cells (1). This causes the victim to become seriously ill and if not treated can result quickly in death (1). Symptoms usually start after about a week to 10 days and can include fever, shaking, chills, sickness, vomiting and muscle pains (1). Children under 5 are often worst affected because they have built up less resistance than adults (1). Malaria can recur and so people may often experience several bouts of illness (1). This has a very serious economic effect on their families as if they cannot work they may lose income (1). As a result, families may not be able to afford to send their children to school, so they lose out on education (1). Their income may be so low that they cannot afford sufficient food and so malnutrition and hunger can also be a problem (1). Crops may be left unharvested in the fields because farm workers are too ill to gather them in (1). The whole economy of a malaria affected country can suffer because of low productivity, as much of the workforce is frequently off sick (1). Few tourists want to visit the country because of the threat from malaria, further hitting the country economically (1).

Or any other valid point.

If **cholera** chosen:

Cholera can cause extreme sickness, vomiting, muscle cramps and diarrhoea within 2 to 5 days of infection (1). This can cause the victim to become dehydrated very quickly due to loss of body fluids (1). This can lead to shock, a severe drop in blood pressure and death if not treated quickly (1). Cholera often has the worst impact on areas where lots of people are living close together in insanitary conditions because the bacteria can spread so quickly from person to person (1). This is often the result of a natural disaster such as an earthquake or hurricane or due to war damage (1). Children are at particular risk and can die from cholera within 24 to 48 hours if they don't receive the right treatment (1). Up to

60% of people who develop cholera will die if they are not treated (1). The impact on communities is therefore very high as workers are off sick and productivity is consequently very low (1). This affects the whole economy of the country as resources are used up fighting the cholera outbreak instead of being invested in other areas such as education (1). People who recover from cholera are often weak and have lowered resistance to fight off other diseases, so their long term health suffers (1).

Or any other valid point.

NATIONAL 5 GEOGRAPHY 2017 SPECIMEN QUESTION PAPER

1. (a) Corrie with lochan in squares 7921 & 8021 (1). Scree slopes on south side of corrie (1). U-shaped valley at 8414 (1). Pyramidal peak (Pen-y-Fan) at 012215 (1). Arete at 016213 (1). Corrie with lochan (Llyn Cwn Llwch) at 0022 (1).

   Or any other valid point.

   (b) Glacier forms in corrie/north facing slope and moves downhill due to gravity (1). Eroding sides and bottom of valley (1) through plucking and abrasion (1). Action makes valley sides steeper and valley deeper (1).

   When glacier retreats a deep, steep, flat floored U-shaped valley left behind (1). Original river in valley now seems too small for wider valley and is known as misfit stream (1).

   Or any other valid point.

2. (a) From point 905175 river is flowing south (1) down steep-sided V-shaped valley (squares 9017 & 9016) (1). River is joined by tributaries from west such as at 906166 (1). Confluence at 911153 (1) and from this point river gets wider (1). Meander at 911152 (1). At least 3 waterfalls marked in square 9009 (1). In this square river is flowing south-west (1).

   Or any other valid point.

   (b) At river meander, water pushed towards outside of bend causing erosion (1), by processes such as corrosion or hydraulic action (1).

   Slower flow of water on inside bend causes deposition (1). Over time erosion narrows neck of meander (1). In time, usually during a flood, river will cut right through neck (1). Fastest current is now in centre of river and deposition occurs next to banks (1) eventually blocking off meander to leave ox-bow lake (1).

   Or any other valid point.

3. Answers will vary depending upon the land uses chosen.

   Tourism and recreation examples: Tourists able to visit show caves at Dan-yr-Ogof in 8316 (1) other attractions nearby which might be of interest such as Shire Horse Centre and public house in 8416 (1). Camp site in this square where would be able to stay (1). Lots of opportunities for outdoor enthusiasts such as walking Brecon Beacons Way (7922) (1) nature reserves such as in 8615 where visitors may see rare wildlife species (1).

   Climbers could tackle the cliffs in Pen-y-Fan (0121).

   Farming: Lots of mountainous land suitable for hill sheep farming (1) sheep able to survive in the colder, windy and wet conditions (1). Farms such as Coed Cae Ddu in 9510 have good road connections, only about 2 kilometres from an A class road, giving good access to markets (1) patches of woodland which provide shelter for the ewes, especially at lambing time (1). Other farms such as Pwllcoediog (8416) able to benefit from high numbers of visitors by earning extra income from bed and breakfast (1).

Industry: Limestone areas such as Brecon Beacons are sometimes used for extraction of limestone (1). Quarries could be built here as there is limestone and an A class road (A4067) nearby for material to be transported (1). Opencast working shows evidence of industry, grid reference 8211 (1). Works located at 847107 where land is flat so easy to build on (1). Good communication with two main roads A4109 and A4221 close by, also rail link into works (1). Small settlements close by like Dyffryn Cellwen where workers could be found (1).

4. (a) Cape Wrath has north wind whereas Banbury has west wind (1). 35 knots at Cape Wrath but calmer in Banbury at 15 knots (1). Dry in Banbury but snow showers at Cape Wrath (1). 6 oktas cloud cover at Cape Wrath but only 2 oktas over Banbury (1). Temperature at Cape Wrath is much colder at 2C, while at Banbury is 11C (1).

   (b) Should not go walking in hills as cold front about to arrive in area (1) which will bring heavy rain showers (1). Will also cause temperature to drop close to freezing point and could be snow (2). If not properly equipped could suffer from the cold and get hypothermia (1) especially as isobars are close together resulting in high wind chill (1). If heavy snow or low cloud they could lose their way easily and need to be rescued (1) these conditions are life threatening and should wait for a better day (1).

   Or any other valid point.

5. Tropical continental air mass will bring hot dry weather in summer which could result in droughts (1). Might need to be hosepipe bans (1). Grass might wither and die causing problems for livestock farmers (1). Ice cream sales might rise (1) as people make most of sunny weather and head for beach (1). Could be very hot and difficult to do physical work outside (1).

   Heavy rain from thunderstorms might cause flash floods (1).

   Or any other valid point.

6. Answers will vary depending upon the land uses chosen.

   Examples of problems between tourists and farmers
   In Cairngorms, tourists can disrupt farming activities as walkers leave gates open, allowing animals to escape (1). Tourists' dogs can worry sheep if let off the lead (1). Stone walls are damaged by people climbing over instead of using gates/stiles (1). Solved by putting in kissing gates or stiles so that people don't have to open gates (1). Noisy tourists can disturb sheep especially during breeding season (1). Farmers may restrict walkers' access at certain times eg lambing season (1). Farm vehicles can slow up tourist traffic on roads (1) parked cars on narrow country roads can restrict movement of large farm vehicles (1). Some of these problems can be resolved by educating public through methods such as publicising Country Code (1).

   Examples of problems between industry and tourists
   Tourists want to see beautiful and unusual scenery of Yorkshire Dales but quarries spoil natural beauty of landscape (1). Lorries used to remove stone endanger wildlife and put visitors off returning to area (1). This threatens local tourist-related jobs eg in local restaurants (1). Large lorries needed to remove

quarried stone cause air pollution and dust which spoils atmosphere for tourists (1). Quarry companies have covered vehicles with tarpaulins to try and reduce amount of dust (1). Lorries cause traffic congestion on narrow country roads which slows traffic and delays drivers (1). Some quarries have reduced number of lorries by sending limestone by rail instead (1). Peace and quiet for visitors is disturbed by blasting of rock (1).

Quarry companies limit frequency and times of blasting to try to reduce impact on local communities (1). Some wildlife habitats may also be disturbed by removal of rock (1).

Or any other valid point.

7. (a) Main roads lead into this square (1) there is a bus station (1) and two railway stations (1) tourist information centre (1) several churches (1) museum (1).

   Or any other valid point.

   (b) Land is flat so easy to build on (1) space available for expansion (1) eg expansion of motor works at 163823 (1). Good transports links like M42 allowing people and products access to and from area (1). Rail link with Birmingham International rail station gives easy access to airport (1).

   Many road junctions and intersections connecting area to other areas and less traffic congestion as is away from Birmingham city centre (2).

   Land is on edge of Birmingham so will be cheaper encouraging housing estates like Sheldon to be built (1). Cheaper land allows houses to be bigger with cul-de-sacs, gardens etc. (1) houses can provide a source of labour for the airport, motor works and business park (1).

   Or any other valid point.

   (c) **Area Y:** Some quieter roads such as cul-de-sacs which would be peaceful area to live in (1). Balsall Heath very close to cricket ground (0684) which would be good for recreation (1) and Cannon Hill Country Park (0683). also close by which would give family easy access to pleasant place to walk (1). Children might also enjoy seeing wildlife at Nature Centre in park (1). Area Y is close to centre of Birmingham which might make it easier for parents to get to work if their jobs are there (1) will also be convenient for shopping as lots of variety in CBD (1). Appears to be industrial areas in Area Y (8410) which might provide job opportunities for parents, conveniently close to where they will live (1).

   Area Z: Further out of city, land prices will be cheaper and may be able to afford better house (1). Evidence of lots of modern housing estates with curvilinear road patterns which will be nice environment to live in as will be more garden space (1) and less traffic making it safer for families (1). Two schools in Area Z which mean children will not have long journey to get there (1). 2 golf course in Area Z, providing recreation opportunities for Russell family (1) and able to get outdoor exercise easily by walking along North Worcestershire Path which passes through Area Z (1). Good transport links into Birmingham for shopping or jobs via main A435 road which leads into CBD and also nearby Park & Ride (1077) next to station where they could travel by train into centre (1).

   Lots of open space and areas of woodland make the environment/air quality better than many other parts of city (1).

   Or any other valid point.

8. Pesticides reduce disease producing better crops (1) and surplus to trade (1). Fertilisers increase crop yields (1) leads to better profits for some farmers (1) which can lead to increase in standard of living (1).

Mechanisation means less strenuous work for farmer (1) and is quicker and more efficient (1). GM crops produce greater yield and are disease resistant so make a greater profit for farmer (1) can reduce cost to farmer of applying pesticides (1) and reduce risk to his health (1).

Growing demand for biofuels means higher crop prices and can result in farmer getting higher income (1) and create employment (1).

Or any other valid point.

9. Stage 2 birth rate was still very high, whereas death rate fell quite quickly — caused rapid rise in total population as people were living longer (1). Death rates fell because clean water supplies introduced, reducing spread of disease (1) and at same time proper sewage systems being built which meant water supplies no longer contaminated, reducing number of people falling ill and dying (2). Advances in medicine such as introduction of penicillin helped keep death rates low, as people could be treated for and cured from illnesses which might have killed them in past (2). Birth rates were still high as people were used to idea many children may not survive until adulthood (1) and also children were required to go out and work because of poverty (1). Stage 3 birth rates started to drop much faster as people realised infant mortality was falling and no longer needed to have extra children as insurance policy (2). Standard of living had improved and wasn't necessary to have lots of children to earn income for family any more (1). Also education about family planning was more common (1) and availability and variety of different methods of contraception was better (1). Falling birth rate meant that rate of population increase started to slow down (1). Stage 4 birth rate fallen so low that in some countries is below death rate and so overall population is falling (1).

Japan is example of this (1).

Or any other valid point.

10. **Life expectancy:** Very useful development indicator as shows that people in developed countries such as Finland live much longer than in developing countries such as Chad (1). Likely to be because standard of living in Finland is much better (1) and will be much better hospitals, more doctors and more money to spend on medicine (1) as Finland is wealthy developed country which can afford to pay for all of this (1). In Chad, people will have very hard physically demanding lives which may lead to shorter life expectancy (1). They may also live shorter lives on average because of poor nutrition, food shortages and famine (1).

Percentage of workforce employed in agriculture: Can tell you a lot about a country because if a very high proportion of workforce in agriculture shows less developed country (1). Whereas in developing countries such as Mali, 80% of workers are employed in farming because mostly subsistence farmers who have to grow own food (1). Few other places that can get food from or simply can't afford to buy it (2). Also few other industries for people to get jobs in as country is less developed and there is lack of money to invest in setting up new businesses (2). Developed countries such as the Netherlands have very efficient farming industries which require very few workers (1); their economy is highly developed meaning most people are employed in many other jobs and industries which are available and which provide higher incomes than farming (2).

Or any other valid point.

11. (a) Overall trend is that amount of Arctic Sea ice has decreased between 1979 and 2013 (1) from (around) 7 million square km to (about) 5 million square km (1). Fluctuation in extent of sea ice in certain years (1) eg amount of sea ice increased from 3.75 million square km in 2012 to 5 million square km in 2013 (1). Between 2006 and 2007 was a sharp decrease (1) from 6 million square km to 4.25 million square km (1).

Or any other valid point.

(b) Increased temperatures causing ice caps to melt so Polar habitats beginning to disappear (1). Melting ice causes sea levels to rise (1) threatening coastal settlements (1). Increase in sea temperatures causes water to expand, compounding problem of flooding (1). Global warming could also affect weather patterns, leading to more droughts (1) crop failures and problems with food supply (1). Flooding, causing the extinction of species (1) and more extreme weather, eg tropical storms (1). Tourism problems will increase as will be less snow in some mountain resorts (1). Global warming could threaten development of developing countries as restrictions on fossil fuel use may be imposed to slow rate of increasing CO_2 levels (1). In UK, tropical diseases like malaria may spread as temperatures rise (1). Plant growth will be affected and some species will thrive in previously unsuitable areas (1). Higher temperatures may cause water shortages (1).

Or any other valid point.

12. (a) High rates of deforestation occur in Brazil, DR Congo and Indonesia (1).

High rates are also prevalent in areas such as Mexico and most of South America (1). High levels of loss more common in developing countries (1).

Moderate levels common throughout Europe, northern Africa and Canada (1). Low rates common throughout USA, China, India and Australia (1).

Or any other valid point.

(b) Management strategies include: Habitat Conservation Programmes sometimes established in tundra environments to protect unique home for tundra wildlife (1). In Canada and Russia, many tundra areas protected through national Biodiversity Action Plan (BAP) (1). BAP is internationally recognised programme designed to protect and restore threatened species and habitats (1). Reducing global warming is crucial to protecting tundra environment because heating up of Arctic areas is threatening existence of environment (1). Most governments have promised to reduce greenhouse gases by signing up to Kyoto Protocol (1). Many countries have invested heavily in alternative sources of energy such as wind, wave and solar power. These sources of energy are renewable and more environmentally friendly than burning fossil fuels, which increase carbon emissions and global warming (2). Some oil companies now schedule construction projects for the winter season to reduce environmental impact (1). Projects work from ice roads, which are built after ground is frozen and snow covered. This limits damage to sensitive tundra (1). Some oil companies locate polar bear dens using infrared scanners and do not work within 1.6 kilometres of these dens (1). Number of Arctic research programmes, such as International Association of Oil & Gas Producers' joint industry programme on Arctic oil spill response technology (1). This programme attempts to increase effectiveness of dispersants in Arctic waters, oil spill modeling in ice and use of remote sensors above and under water (2). Many companies operate sophisticated systems to detect leaks (1). Many companies work with local communities to understand and manage potential local impacts of their work (1). Many countries have set up national parks such as Arctic National Wildlife refuge in Alaska to protect endangered animals in tundra (1). Trans-Alaskan pipeline is raised up on stilts to allow Caribou to migrate underneath (1).

Or any other valid point.

13. (a) Over last 100 years number of eruptions has increased from forty three in 1910s to seventy in 2010s (1). Apart from decades of 1920s, 1970s and 1990s, amount of volcanic activity in each decade increased (1).

Least number of eruptions was in 1920s with only 31 (1). Big drop between 1910s and 1920s with a drop of 12 eruptions (1). Also in 1990s there were 12 fewer eruptions than 1980s (1). Biggest increase between 1990s and 2000s with 13 more eruptions (1). Decades with greatest number of eruptions were 1980s, 2000s and 2010s at 66, 67 and 70 (1).

Or any other valid point.

(b) For Pico de Fogo volcano

Heat from lava flows set fire to main settlements destroying two villages as well as forest reserve (2) endangering the vegetation and animal habitat (1). Around 1,500 people forced to abandon homes before lava flow reached villages of Portela and Bangeira on Fogo island (1). More than 1,000 people evacuated from Cha das Caldeiras region at foot of volcano to ensure safety and prevent injuries (1). Airport was closed, as ash filled sky, to prevent risk of planes crashing (1). Buildings and records were destroyed resulting in some of history of area being lost (1). Roads and transport routes destroyed, affecting tourist industry on island (1). Volcano destroyed agricultural land which resulted in loss of fertile land (1) decreasing ability of area to produce crops (1) and support local population (1). Tourism might increase as volcano becomes tourist attraction improving economy of the island (1).

Or any other valid point.

14. (a) Europe dominated world trade exports with around 43% in 2005 (1).

Dropped to around 38% in 2010 (1). Europe still largest exporter in 2010 (1). Asia had second largest regional share of world trade with around 27% in 2005 (1), growing to around 31% in 2010 (1). Africa's share is low, around 3%, (1) but has grown by about 1% (1). North America's share has dropped from just under 15% in 2005 to around 14% in 2010 (1).

Or any other valid point.

(b) Farmers paid fair wage for their work (1) and safer working conditions promoted (1). Money from fair trade can be used to improve services in local communities (1) such as schools and clinics (1) which improves standard of living (1). More money goes directly to farmer, cuts out middlemen who take some of profits for themselves (1). Farmers receive guaranteed minimum price so are not affected as much by price fluctuations (1). Fair trade encourages farmers to treat workers well and to look after environment (1). Often fair trade farmers are also organic farmers who do not use chemicals on crops so protect environment (1). Health care services and education programmes available and tackle problems of HIV/AIDS(1).

Or any other valid point.

15. (a) USA has six out of ten most popular tourist attractions in world including Niagara Falls and Disneyland (1). Most visited tourist attraction is Times Square in USA with 35 million visitors per year (1). Washington D.C. is second most popular tourist destination with 25 million visitors (1). Trafalgar Square is most popular tourist area in Europe (1). Notre Dame and Disneyland in Paris are most visited attractions in France with 12 million and 10.6 million visitors a year (2). Disneyland Tokyo is most visited attraction in Asia (1). Four out of top ten most popular tourist destinations are Disneyland/Disneyworld parks located on 3 different continents (1).

Or any other valid point.

(b) **People (positive):** Local people employed to build tourist facilities eg hotels (1) and work in restaurants and souvenir shops (1). Employment opportunities allow locals to learn new skills (1) eg obtain foreign language (1) and earn money to improve standard of living (1). Services improved and locals can benefit by using tourist facilities such as restaurants and water parks (1). Better employment opportunities increase the local Governments' revenue as wages are taxed (1) so can invest in schools, healthcare and other social services (1). Locals can experience foreign languages and different cultures (1) and can benefit from improvements in infrastructure eg roads and airports (1).

People (negative): Tourist-related jobs are usually seasonal therefore some people may not have income for several months (1) eg at beach and ski resorts (1). Large numbers of tourists can increase noise pollution and upset peace and quiet (1). Local people may not be able to afford tourist facilities as visitor prices are often higher than local rates (1). Tourists can conflict with local people due to different cultures and beliefs (1). There is additional sewage from visitors which increases risk of diseases like typhoid and hepatitis (2).

Environment (positive): Appearance of some areas can be improved by modern tourist facilities (1). Some tourists are environmentally conscious and can have positive impact on landscape by donating money to local projects which help protect local wildlife (1) eg nature reserves (1).

Tourist beaches cleaned up to ensure safe for people to use (1) through initiatives like Blue Flag (1). Seas become less polluted as more sewage treatment plants built to improve water quality (1).

Environment (negative): Land lost from traditional uses such as farming and replaced by tourist developments (1). Traditional landscapes/villages spoiled by large tourist complexes (1). Air travel increases carbon dioxide emissions and contributes to global warming (1). Traffic congestion on local roads increases air and noise pollution (1). Tourist facilities such as large high-rise hotels and waterparks spoil look of natural environment (1). Litter causes visual pollution (1). Increased sewage from tourists can cause water pollution (1). Polluted water damages aquatic life and habitats (1).

Or any other valid point.

16. (a) In April 2014 were few cases of Ebola in Africa. By October 2014 were almost 2500 cases in Liberia (1). In Sierra Leone were almost 1,200 cases by October 2014 (1). In Guinea were around 800 cases by October 2014 (1). In Liberia cases rose rapidly from around 250 in August 2014 to around 2500 by October 2014 (1). Sierra Leone witnessed rapid increase in cases from around 500 cases on October 1st 2014 to almost 1200 by mid October 2014 (1).

Or any other valid point.

(b) **Heart Disease** Lifestyle factors are main cause of heart disease. Many people do not take enough physical exercise which is necessary to keep heart healthy (1).

In developed societies many people take car or use lift rather than walking/taking stairs (1). Poor diet leads to heart disease (1). Too much saturated fat can cause hardening or blocking of arteries (1). Many people do not eat enough fruit or vegetables, this can contribute to heart disease (1). Eating too much processed food, with high salt content can also contribute to heart disease (1). Smoking can increase risk of heart disease (1). High stress levels also contribute to heart disease (1). Possible effects of hereditary factors (1).

Cancer
Unhealthy lifestyle is root cause of about third of all cancers (1). Smoking causes almost all lung cancer (1). Poor diet has been linked to bowel cancer, pancreatic cancer and oesophageal cancer (1). Heavy drinking may be a factor in development of cancer (1). Some people may be genetically predisposed to some cancers, eg breast cancer (1). Too much exposure to sun can cause skin cancer (1). Obesity has also been linked with increased cancer risk (1).

Asthma
Infections such as colds or flu affect lungs and narrow airways, making asthma worse (1). Allergic reactions to dust mites in home can cause asthma (1). Pollen from plants outside can cause asthma (1). Traffic fumes in polluted towns and cities can cause asthma (1). Cigarette smoke can cause asthma (1). Asthma can be caused or made worse by damp conditions in home (1). In cases of severe dampness, mould spores may make asthma worse (1).

Or any other valid point.